Winning at Technological Innovation

McGraw-Hill Engineering and Technology Management Series
Michael K. Badawy, Ph.D., Editor in Chief

RITZ • *Total Engineering Project Management*
STEELE • *Managing Technology*
WHEELER • *Computers and Engineering Management*

In memory of my mother
Mrs. Blanche M. Englert

Contents

Series Introduction

Technology is a key resource of profound importance for corporate profitability and growth. It also has enormous significance for the well-being of national economies as well as international competitiveness. Effective management of technology links engineering, science, and management disciplines to address the issues involved in the planning, development, and implementation of technological capabilities to shape and accomplish the strategic and operational objectives of an organization.

Management of technology involves the handling of technical activities in a broad spectrum of functional areas including basic research; applied research; development; design; construction, manufacturing, or operations; testing; maintenance; and technology transfer. In this sense, the concept of technology management is quite broad, since it covers not only R&D but also the management of product and process technologies. Viewed from that perspective, the management of technology is actually the practice of integrating technology strategy with business strategy in the company. This integration requires the deliberate coordination of the research, production, and service functions with the marketing, finance, and human resource functions of the firm.

That task calls for new managerial skills, techniques, styles, and ways of thinking. Providing executives, managers, and technical professionals with a systematic source of information to enable them to develop their knowledge and skills in managing technology is the challenge undertaken by this book series. The series will embody concise and practical treatments of specific topics within the broad area of engineering and technology management. The primary aim of the series is to provide a set of principles, concepts, tools, and techniques for those who wish to enhance their managerial skills and realize their potentials.

The series will provide readers with the information they must have and the skills they must acquire in order to sharpen their managerial performance and advance their careers. Authors contributing to the series are carefully selected for their expertise and experience. Although the series books will vary in subject matter as well as approach, one major feature will be common to all of them: a blend of practical applications and hands-on techniques supported by sound research and relevant theory.

The target audience for the series is quite broad. It includes engineers, scientists, and other technical professionals making the transition to management; entrepreneurs; technical managers and supervisors; upper-level executives; directors of engineering; people in R&D and other technology-related activities; corporate technical development managers and executives; continuing management education specialists; and students in engineering and technology management programs and related fields.

We hope that this series will become a primary source of information on the management of technology for practitioners, researchers, consultants, and students, and that it will help them become better managers and pursue the most rewarding professional careers.

Michael K. Badawy

PROFESSOR OF MANAGEMENT OF TECHNOLOGY
THE R. B. PAMPLIN COLLEGE OF BUSINESS
VIRGINIA POLYTECHNIC INSTITUTE AND STATE UNIVERSITY
FALLS CHURCH, VIRGINIA

Preface

All of us should understand the importance of technological innovation, not only for a company but also for the nation—it is a survival mechanism. As such, it permits businesses to adapt to environmental changes.[1] This adaptability leads to the tangible results of research and development (R&D)—company profits, growth, and leadership—and to such intangible benefits as enhancement of a company's image and morale. Examples of major "survival" innovations are the laser, the microprocessor, and the pacemaker. Such developments have had a profound impact on our economy, our technical leadership as a nation, and our general welfare.

Innovate and survive.

Innovation also contributes to our potential for national growth,[2,3,4] and, as reported by the Industrial Research Institute,[5] technological innovation promotes growth of GNP. Gilpin[6] refers to Robert Solow who, in 1957, related national and industrial productivity to advances in science and engineering. Sekulow[7] says that innovation stimulates productivity and investment which in turn stimulate innovative activity. It is well documented that firms with good track records of innovations have substantially higher productivities and growth rates than those with fewer innovative activities.[8]

Innovative companies are productive companies.

The Concern Over Innovation

Recently, there has been increasing concern over the decline of innovation in the United States and the unfavorable impact of this decline on our general economic well being. In 1980, the Committee on Technological Policy[8] reported serious signs that the United States was losing its technological lead. Horan[9] has also spoken of the concern among knowledgeable industrial and governmental persons over the decline in U.S. productivity, and he relates this to a decline in innovation.

There is also evidence[10,11,12] of a decline in R&D expenditures as a percentage of GNP since 1965, of a decline in radical innovations, and of a decline in the fraction of U.S. patents issued to U.S. citizens. One reason for this lag in innovation in some industrial sectors is the decrease in the growth rate of industrial investments in long-term R&D programs and in the incorporation of the results of R&D into manufacturing processes.

Washington mismanagement has been claimed to be a roadblock to American research.[13] In addition, such elements as foreign competition, tax structures, capital formation, and the rising costs of labor and materials force corporate attention away from an enthusiastic pursuit of technological innovation. This may move managements to emphasize product improvements and to defer or eliminate many of their long-range research projects.

R&D productivity will gradually decline under such conditions. At that juncture, some managements may believe that just because significant product upgrades have been possible in the past, they should be possible now, and that any decline in R&D productivity is the fault of R&D and engineering, rather than recognizing that maturing technology and product life cycles are facts of life. With the certainty of death and taxes, as a product and its technology mature, there is a decreasing chance for significant change or improvement. In fact, failure to recognize this or a substantial change in the market can be fatal.

Mature technologies do not lead to exciting new products.

During any decline in business, the longer-range R&D is usually the first to be cut, and in any recessionary period, product improvement R&D may be seriously impaired. Although long-range R&D means *survival*, it is usually the short-term, bottom-line viewpoint that wins over the long-term welfare. Companies that adopt this philosophy may protect the immediate bottom line, but they will come out of the recessionary period weak at the marketplace for want of any significant product leadership. Hence, companies must exert every effort to maintain their R&D programs, to maintain a reasonable proportion of new-product and -process research, and to discover and use new technologies if they are to increase their R&D productivity and obtain accompanying tangible and intangible benefits.

The Basis for Enhancing Innovation

In order to enhance innovation by increasing the growth rate of industrial investments in R&D and incorporating more R&D results into manufacturing, we must ascertain on what each depends. Long-range industrial R&D depends on the fund of basic research from our universities or our internal research, and both depend on the predilection of management for support. Getting R&D results into manufacturing, that is, winning at innovation, depends on managers' attitudes toward, and performances in, the innovation process. This book discusses managers' roles in this process.

Few industrial companies conduct basic research, so having new technologies upon which to base entirely new products means that industry must advance its fund of basic research by ensuring that university research has continuing support.[14] Although government laboratories also are sources of basic research, I believe the key to industrial innovation lies in a strong industry and university relationship. Strong linkages are, therefore, necessary.

Basic research spawns new technologies; new technologies lead to industrial innovation.

An industry's support of basic research and the application of the resulting new technologies in applied R&D mean taking greater risks. But if a company has the tough, strong leadership that can handle a reasonable proportion of higher risk R&D, the payout is there. A strong connection exists between new-product sales growth and market share,[15] and return on investment has been shown to increase with market share.[16]

When R&D groups explore new technologies without getting new products based on them into the marketplace, management impatience with R&D will grow, unless it is clearly shown that the flow of R&D results into engineering and manufacturing has been throttled or impaired by the reluctance of an operating group to get involved with substantial change. The basis of such reluctance is fear and the cost of change. Change involves new initial costs at the expense of the immediate bottom line. To overcome such reluctance, the R&D group must initiate collaboration with other groups early, and management must foster such collaboration.

An R&D responsibility is to seek collaboration with other groups.

To offset any decline in a company's productivity and profitability other than just a flat-out cutback in R&D by management, managers at all levels must recognize the need for innovation, foster basic research, and urge a balance of long-range, new-product research in the total R&D project portfolio. This book identifies and discusses those actions by managers that promote getting R&D results through engineering and manufacturing to sales. The results of promoting longer-range R&D and managers' performances will be new profits and growth from an enhanced productivity.

About the Text

This book is designed as a holistic guide to how managers can win at technological innovation. I submit that there are two aspects to winning: being a good technical manager and enhancing your chances of getting R&D results through the innovation process into the marketplace.

The text has three parts: In Part 1, the innovation process and management style are discussed. Part 2 describes the activities necessary to good technical management. Part 3 presents methods and techniques useful to the activities presented in previous chapters. I have simplified the organizational structure in the text by having area managers that represent disciplines and (or) functions in a corporate R&D group reporting to a vice president of R&D.

A reference and bibliography section is at the end of each chapter.

Cases and Incidents about a Fictitious R&D Group

At the end of each chapter are several cases and incidents—events that might, should, or shouldn't happen in an R&D group. They are fictional. Their purpose is both to entertain and to instruct and challenge by provoking your thinking about the events described. Questions at the end of each story are intended to stir your thinking about possible ramifications of incidents in the stories. The cases can also be used for discussion purposes in small groups in management development programs.

The setting for the stories is a small corporate laboratory in the XYZ Company. The CEO formed the lab some 2 1/2 years ago to get into businesses that he felt were foreign to those currently active in operational groups. He had also taken an option on an alloy mixture which he felt offered unique opportunities. The CEO had not considered mission, goals, or operating procedures in any depth but had decided that the lab's initial objective was to explore new high-strength alloys that might go into products that would give the company a leg up on competition.

Remember the setting and the cast.

The cast starts with Dr. Helms, who is the leader of the R&D group. He came to XYZ after several years in a local research institute as a metallurgist. Before that he spent almost five years in a company that manufactured valves. At XYZ he has three areas, each with a manager, four to five professionals, and a like number of technicians. Roy is manager of fundamental alloy investigations and is the key professional behind existing efforts, being an excellent engineer with a background pertinent to the investigation. Larry is manager of applications development and a journeyman engineer; he is creative, and what you would call a nice guy. Tim is manager of analytical investigations. Jay is a professional in Roy's area. Joe is an engineer in Larry's area, and Harry is a senior electronics engineer currently in Larry's area. Hank is an experience-wise technician who is primarily responsible for the shop and equipment maintenance.

Dr. Watts is a consultant who has the confidence of the CEO. The CEO has asked him to look at this group and see why no results have been seen in the marketplace, to see what else the group can do, and to evaluate current efforts and staff.

No new results; no new profits.

As the stories continue, I will tell you of staff changes as they occur and remind you of an individual's position when it may be helpful.

Acknowledgments

I am grateful to many colleagues for their review of and advice regarding specific sections in this book. These include Messrs. John Hazelwood and Daniel Rubin on patents; Drs. Martin Cooper and Wm. S. Birnbaum on planning; Mr. J. R. Jennings on profitability measures; Mr. Ralph Lenz on technical forecasting; Mr. James Haughey on the rate-of-change method of forecasting; Dr. M. M. Menke on decision analysis; Mr. E. D. Porro on market research; Mr. J. E. Gorgens on constraints to commercialization; and Dr. F. G. Sawyer for his general review of the text. Others who gave valuable suggestions include Messrs. W. E. Dean, M. K. Dishman, James Helms, M. J. Kammerer, L. P. Shefferly, Earl Weed, and Dr. M. A. Von Glinow.

Special acknowledgment and thanks are given Mrs. Estelle Doyle for extensive editing of the text and for her many helpful suggestions and to Dr. David A. Rohy for his detailed review of the text and his valuable suggestions.

Robert D. Englert

References

1. *Investments in Tomorrow*, 7(1):3 (1977) SRI International, Menlo Park, Calif.
2. J. Berger, "Productivity: Why It's the No. 1 Underachiever," *Business Week*, April 20, 1987, pp. 54-55.

3. D. M. Keezer, "Research and Development and National Growth," SRI International, LRPS Client Conference, Menlo Park, Calif., October 1967.
4. "The Silent Crisis in R&D," *Business Week*, Mar. 8, 1976.
5. Industrial Research Institute, "The Impact of Industrial Innovation on the Economic and Social Welfare of the U.S.," New York, Oct. 19–22, 1980.
6. R. Gilpin, "Technology, Economic Growth, and International Competitiveness," *A Report to the Subcommittee on Economic Growth*, U.S. Government Printing Office, Washington, D.C., July 9, 1975.
7. E. Sekulow, "Is America Losing Its Technological Leadership?" Pub. Rel. Soc. of Amer., Detroit, Mich., Feb. 26, 1980.
8. T. H. Vanderslice, "Stimulating Technological Progress," Committee for Economic Development, New York, January 1980.
9. J. J. Horan, "Innovation: Key to the Future," Annual Meeting of the Pharmaceutical Manufacturing Association, excerpts by Merck & Co., Rahway, N.J., 1980.
10. W. S. Abernathy, "Competitive Decline in U.S. Innovation: The Management Factor," *Res. Mgmt.*, September 1982, pp. 34–41.
11. National Science Foundation, "The 5-Year Outlook on Science and Technology—1981," 2 vols., U.S. Government Printing Office, Washington, D.C., 1982.
12. S. Kindel, "What's Wrong With Corporate R&D," *Financial World*, Jan. 26, 1988, pp. 24–30.
13. S. Siwolop, "Technology," pp. 38–41; R. Wrubel, "Washington Hits and Misses," pp. 43–44; and A. Baldo, "Wish List," pp. 46–48, *Financial World*, Nov. 14, 1989.
14. R. Barker, "Bringing Science into Industry from Universities," *Res. Mgmt.*, November–December, 1985, pp. 22–24.
15. The Strategic Planning Institute, "New Products and Market Position," *The Pimsletter on Business Strategy*, no. 28, Cambridge, Mass., 1982.
16. The Strategic Planning Institute, "The Dispute about High-Share Businesses," *The Pimsletter on Business Strategy*, no. 19, Cambridge, Mass., 1979.

Suggested Readings

Abetti, P. A.: "Innovation from Start to Finish, "*Chemtech*, July 1980, pp. 405–412.

Adler, S. F., and H. H. H. Fang: "U.S. Patent Productivity," *Res. Mgmt.*, September-October, 1986, pp. 29–35.

Hopkins, D. S.: "New Product Winners and Losers," *Res. Mgmt.*, May 1981, pp. 12–17; also in *The Conference Board Report* no. 773, 1980.

———, and E. L. Bailey: "New-Product Pressures," *The Conference Board Report*, June 1971, pp. 16–24.

Hughes Aircraft Co.: *R&D Productivity*, 2d ed., Culver City, Calif., 1978.

Industrial Research Institute: "Innovation within the Corporation," New York, Mar. 30, 1979.

———: "Living Case Histories of Industrial Innovation: Selected Papers from *Research Management*," 1973–1980, New York, 1981.

Mansfield, E.: "Basic Research and Productivity Increase in Manufacturing," *The Amer. Econ. Rev.*, December 1980, pp. 863–873.

Nason, J. K.: "The Environment for Industrial Innovation in the United States," in *Technological Innovation: Government/Industry Cooperation*, A. Gerstenfeld and R. Brainard, Wiley, New York, 1979.

Sadler, G. E.: "Productivity Perspectives," American Productivity Center, Houston, Tex., 1981.

Souder, W. E.: "Encouraging Entrepreneurship in Large Corporations," *Res. Mgmt.*, May 1981, pp. 18–22.

Tushman, M. L., and W. L. Moore: *Readings in the Management of Innovation*, 2d ed., Pitman, Boston, 1988.

Part

1

The Innovation Process and Management Style

We must first establish a mutual understanding of the innovation process, and then I will discuss those elements required for a winning operation.

Technical innovation is described as a six-phase process. In phase 1, viable concepts are created. Research on the concepts' feasibility is done in phase 2, followed by product and process research and development in phase 3. In phase 4, the manufacturing prototype is produced. The last phases involve manufacturing and sales. (See Chap. 1.)

Management style is discussed as a basis for influencing others involved in the innovation process. For our purposes here, style *is the way a manager establishes, maintains, and nurtures relationships that enhance productivity of the innovation process. The basis for style is presented, including managers' characteristics, skills, and approaches to management. The environment of the R&D manager is described, and a manager's internal environment is illustrated as "innovation's management circle." The R&D manager is shown in this circle as interacting with the boss; the staff; the corporate operations vice presidents; the division presidents; and the downstream engineering, manufacturing, and marketing managers. This leads to a discussion of the rapport the R&D manager must promote with staff and the several managers represented. (See Chap. 2.)*

Chapter

1

The Innovation Process

Questions answered in Chapter 1

- What are the different ways companies acquire R&D results?
- What are the six phases of the innovation process?

Companies grow through their own or others' innovations: We do R&D, we buy or license others' R&D, we acquire companies and their R&D, or we join other companies in ventures of mutual interest. Our interest here is in-house technical innovation—the creation of an idea by our personnel and its development to a product or process that sells in the marketplace. Innovation is not just an activity assigned to an R&D group. Technological innovation is a process, and to be successful it requires the companywide application of all scientific and technological talent available as well as such related activities as marketing. By this process, we convert resources into results, that is, into profits, growth, and leadership. This process is depicted in Fig. 1.1.

R&D is but a small part of the total innovation process.

The process of technological innovation has six phases and can be under the direction and responsibility of several managers of different groups before it spawns a product for a consumer (see Fig. 1.2). In phases 1 through 3, the R&D group selects and develops ideas in a way that promotes or negates success in the follow-on tasks—phases 4 through 6. In phase 1, the group needs many ideas in order to arrive at a few outstanding concepts. Activities that promote creativity (such as brainstorming) are used.

R&D is concerned primarily with creation and development.

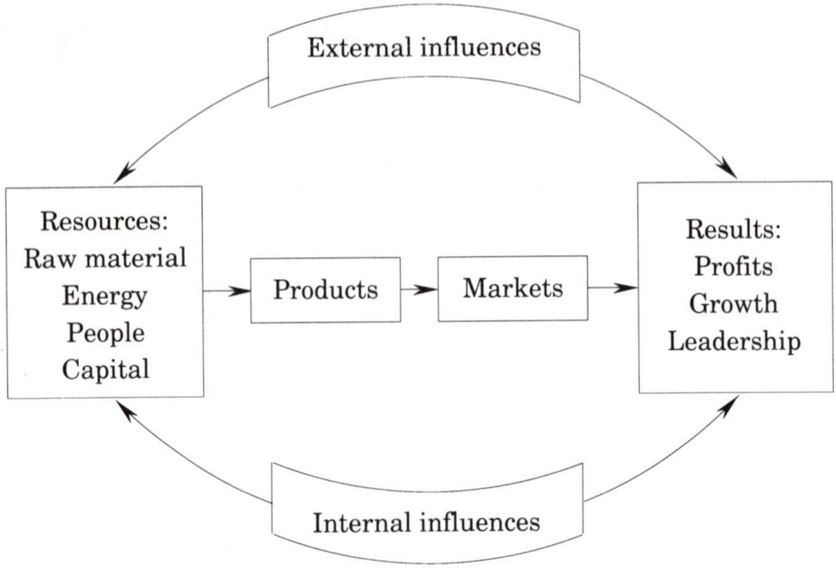

Figure 1.1 Transforming resources into results.

In phase 2, research on the concept's feasibility, the group studies technical, legal, and market constraints; conducts the critical experiments (based on the constraints identified) needed to assess technical promise; obtains the data needed to suggest the existence of a reasonable market; and surveys competitive products to assess whether a product advantage may exist.

During phase 3, product and process research and development occur. The R&D group determines optimum conditions and materials and studies the idea in a pilot plant. Additional research provides fundamental data on parameters affecting performance. The group develops engineering costs and compares them with the best data on competitive products or processes. This is also the point at which R&D files patent applications and makes some intelligent guesses about foreign applications. Interest of possible customers is also assessed at this point, taking care to discuss proprietary aspects with outsiders only under confidential agreements that protect foreign patent rights.

In phase 4, engineering produces the manufacturing prototype or puts the works plant through its paces. If a product manager has not yet been appointed, such appointment must occur here. The engineering staff studies manufacturing methods and develops manufacturing costs. They conduct field trials and develop final designs. They select sales methods and incorporate quality control into the total process.

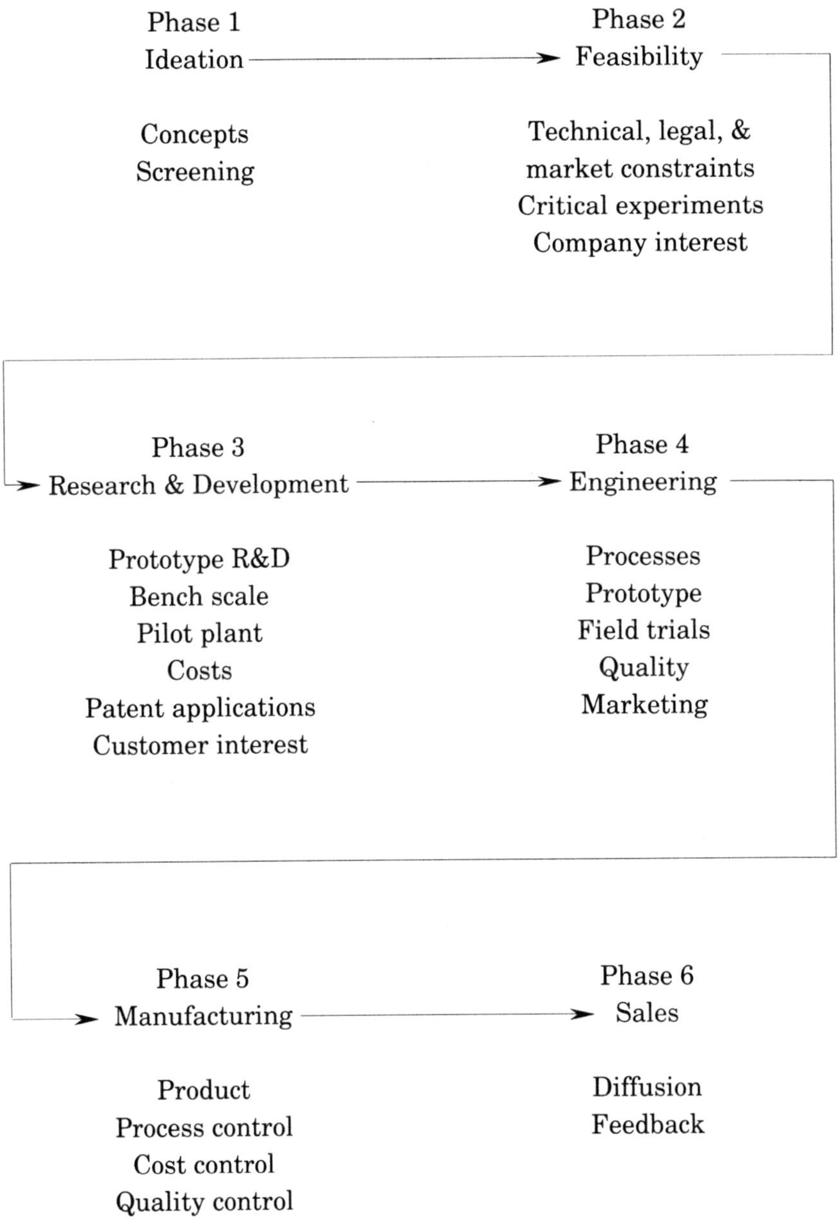

Figure 1.2 The innovation pipeline.

Finally, engineering prepares a business plan and submits it for approval.

Assuming approval, the manufacturing and marketing departments, which have been essential parties in the process, carry on the fifth and sixth phases of the innovation process. Sales in the marketplace are called the *diffusion phase*. In this final phase, manufacturing (with its efficiency of production), sales (with its aggressiveness at the marketplace), and the product's technical performance (in the eyes of the user) have large influences on the degree and facility of diffusion.

In this discussion, I present the R&D group as separate from engineering and manufacturing, but in many instances it is the research and engineering (R&E) group that passes results to manufacturing. In these instances, often there is seldom much research. The setup given here permits a better discussion of the subjects, in my opinion.

A concept from the R&D group must pass through these six phases, and it can get lost in any one of them. An R&D group cannot operate independently of the groups concerned with the other phases. To be successful, persons involved with each successive phase must have the same enthusiasm and dedication exhibited by the R&D group. This means that the R&D group must obtain early participation of the downstream groups, and the total process must be supported and integrated by management. Thus, successful innovation is linked to how well managers perform their respective roles.

Although seemingly the responsibility of the R&D manager, innovation is the responsibility of managers in several groups.

Cases and Incidents

He Marches to a Different Drummer

Helms is sitting in his office trying to figure out the best way to "get on top of things" when Roy comes charging in. Roy is the personification of neatness. In his office, everything is filed, labeled, cross-indexed, and buttoned down. To Roy, litter and sin are synonymous.

Roy is obviously upset. "I've about had it with Jay. He gets in around 10 almost every morning. I've talked to him about it and he keeps sucking on that damn pipe and mutters, "Uh huh, uh huh." His office is a catastrophe. He has reports, equipment, books, reprints everywhere but on his chair! He never throws *anything* away!"

Helms, who knows all about Jay's eccentric habits and also

knows he is an excellent professional, asks whether Jay stays late. Roy says that he usually stays until 6:30 or 7:00.

Helms asks if Jay can find a paper or report when Roy asks for it. Roy nods: "It's amazing! He will suck on that damn pipe for a moment and then lift the report out of one of the stacks."

Helms then asks the crucial question: "What would you do if he left or you fired him?"

Roy looks baffled for a moment and replies disgustedly, "I'd have to close out two of our projects. One is based on his idea, and he's the only one who understands the math; the other depends on how he leans on the damn equipment!"

Helms laughs. "My suggestion to you is that you relax and ride with it. No one is going to bitch about his hours if they can't fill his shoes. And you can always close the door on his office. We all have our differences, you know, and you must recognize that fact. If your man marches to a different drummer, ask yourself if you want him, and if you do, leave him in the environment in which he works best."

Helms beats a small tattoo on his desk as Roy leaves.

Questions

- Rules are rules, or are they?
- Would you have told Roy something different?
- What about safety?
- What effect does a person's moving the 8-hour day forward a couple of hours have on other staff?
- If Roy were to get Jay to adhere to his concept of an 8-to-5 day, do you think Jay's productivity would go up or down? Why?

Ethics (Or What Will You Do When Tempted?)

Helms asks Watts what he thinks about professional ethics. Watts replies, "Well, I have rather rigid standards. I hope they are high. I would be damned upset to be accused of dishonesty. But many people I've known have pretty sloppy ethical standards, in my opinion, and don't think of themselves as dishonest. They have a viewpoint that flexes on occasion.

"For example, there was the biochemist caught graphiting results. Falsified records advance neither the science nor the individual. And there was the manager who misrepresented a project concept to a superior, with the result that the project was canceled. He then told his staff that he did his best with his 'holiness' but the boss unfairly rejected the idea. And the chemist who, when terminated for cause, rifled the group's personnel file for her jacket. She didn't know that duplicate sets were kept!

"There was also the case of the consultant who brought two parties together who didn't know that the consultant was collecting from both of them. I haven't heard from him lately. I also heard of a senior organic chemist who wrote a paper on a fundamental finding originally perceived by a junior chemist, without mention of the junior chemist. And the investigator who purposely underestimated the cost of R&D in order to sucker in support by management.

"I would find it hard to live with myself if my actions were not ethical. But others sleep easily. Perhaps they have no ethics, can't see the long-term result of such actions. Most professionals have standards, although some let certain circumstances seduce them into overlooking (or bending) them sometimes. It's interesting that, in the long run, ethical integrity keeps you out of a lot of trouble and gains you a lot of respect."

Helms nods: "Amen!"

Questions

- Have you read any technical associations' statements on ethics?
- Are the actions in the examples repugnant to you?
- If you found a colleague committing a breach of ethics, what would you do? How about talking to your superior? What if he or she is the culprit?
- Watts appears to be Helm's mentor in lieu of one in the company. Do you have a mentor, preferably within the company, with whom you could discuss such a case?

Suggested Readings

Collier, D. W., J. Monz, and J. Coulen: "How Effective Is Technological Innovation?," *Res. Mgmt.*, September–October 1984, pp. 11–16.

Fowler, D. R.: "University-Industry Research Relationships," *Res. Mgmt.*, January–February 1984, pp 35–41.

Mendell, S., and D. M. Ennis: "Looking at Innovation Strategies," *Res. Mgmt.*, May–June 1985, pp. 33–39.

Roberts, E. B.: "Managing the Process of Invention and Innovation," *Res. Mgmt.*, January 1988, pp. 11–29.

Chapter

2

Management Style and the R&D Manager's Environment

Questions answered in Chapter 2

- What are the components of management style?
- What are the characteristics of a good manager?
- What are the skills of a good manager?
- What are the various approaches to managing people?
- What environmental associations must a manager nurture?
- Who makes up innovation's management circle?
- How do you influence members of this network of managers?

R&D managers want to control and direct the innovation process successfully. My experience supports the conclusion that an R&D manager's success is not only a function of how well he or she may plan, implement plans, lead, and control, but also a function of how good his or her management style is, especially in relating to others involved in the innovation process. An innovation effort that depends on a manager's singular talent, such as technical expertise or an ebullient personality, seldom makes it to the market. Basic to a manager's activities is the environment in which he or she must function effectively.

Style + performance = results.

An Overview of Management Style

It is style which complements affirmation
with limitation and with humility; it is style

*which makes it possible to act effectively, but not absolutely; it is style which enables us to find harmony between the pursuit of ends essential to us, and the regard for the views, the sensibilities, the aspirations of others; it is style which is the difference that action pays to uncertainty; it is above all style through which power defers to reason.**

J. ROBERT OPPENHEIMER

As stated above, success in innovation depends on top-caliber people, working together to accomplish goals in an effective and efficient manner. Such a situation requires that managers have a management style that, first, fosters innovation and a supportive corporate climate and, second, recognizes the need for a totality of effort among all managers.

Management style starts at the top and becomes part of the corporate image and culture. When you think of 3M, what is your mental image? Entrepreneurship. When you think of the oil service industry, what company comes to mind as an aggressive marketer and innovator? Schlumberger. These are managements that have style! All companies will gain an image—good, bad, or nondescript—depending on top management actions.

Corporate culture reflects all actions of management.

I will be using some words that warrant clarification—*climate, culture, environment*—and I have derived definitions from my own usage with the assistance of a good discussion of these subjects by Miller.[17] *Environment* is the atmosphere created by the effects of a company's philosophies, policies, and procedures on employee behavior; it is outwardly evidenced by a particular management style, the type and degree of sophistication of the technology base, the goals set forth, and even the specific plant and equipment. *Climate* is a metaphysical distillate of the behavior of an organization's employees that people sense about the organization. *Culture* is the image projected by the way things are and have been done over a span of years and by the traits established by employees over these years.

Von Glinow[18] has noted that an organizational culture may fall into one of four patterns:

- *Apathetic:* A culture evolving from a lack of concern for people and an indifference about their performance

*L. J. Peter, *Peter's Quotations,* William Morrow, New York, 1977.

- *Caring:* A culture in which there is a high concern for people but few demands on performance
- *Exacting:* A culture in which there is little sensitivity to people but many demands on performance
- *Integrative:* A culture in which there is both a high concern for people and a strong performance expectation

I suggest that we side with the integrative culture.

On a corporate basis, management style and the corporate climate and culture reflect many things: how well top management fosters innovation; whether participation of all functions in innovation is encouraged; how satisfactory the pay and incentive programs are; what type of manager is hired (and thus what type of staff is hired); whether cross-communication is encouraged; whether managers set goals and high standards with the expectation that they will be met; whether creativity is encouraged and recognized; whether the research, development, and engineering budget includes an adequate proportion for true, long-range R&D; and whether management is credible or is just paying lip service to these various elements. We want a corporate management style that has all these elements working together to promote successful innovation. (I have touched only on aspects of direct interest to our subject; other aspects include the company's sales and marketing approach, acquisition orientation, product cost and quality, and so on.)

Having recognized corporate style, let's concentrate on the individual's management style and especially on that of the R&D manager. On an individual basis, *management style* is how a manager behaves and orchestrates his or her activities; it is, in part, the image a manager projects. It is the flair with which a manager promotes the accomplishment of tasks while meeting the needs of coworkers. It is an ability to be firm while being fair.

Your style is reflected in how you conduct your activities.

A manager's style encompasses the entire spectrum of relations between manager and employee, between manager and managers and employees in other company functions, and between manager and superior (remember innovation's management circle). It is the attitude and behavior of a manager as seen by others and as evidenced in results obtained over a period of years. Almost anybody can look good in the short term; in the long term, effective style, as Oppenheimer says, "is style which enables us to find harmony between the pursuit of ends essential to us, and the regard for the views, the sensibilities, the aspirations of others." In other words, if you wish to influence people over time, you will have empathy for their needs as well as concern for your own and the company's needs.

A good manager has a style that influences people favorably.

On both a corporate and an individual basis, a management style is dynamic. The literature monitored, the interplay with colleagues, the conferences attended, and accumulated experiences are great teachers and establish the foundation of a receptive manager's style.

The Basis of Style[19–26]

Style has many facets; it comprises certain characteristics, skills, and approaches to managing people. Because each of these has several aspects, no one manager is like another and no one style is quite like another, or even the "best" style. However, some combinations clearly distinguish one style from another and, hence, one manager from another.

Understand people; influence people; obtain results.

What are the characteristics of a good manager? What basic skills must a manager have? Is the person familiar with the different concepts regarding the management of people? In other words, what is the makeup of a good manager, and what does a manager do?

Characteristics of the effective manager

Good managers need to manage. McClelland[27] says that effective managers have a strong need for power (a disciplined need to influence people) rather than for their own technical achievement. However, there are managers who attempt to conduct research while still managing. While this is no great crime (providing this manager manages only one group), problems arise when such a manager is responsible for several additional groups and continues to conduct research with his or her own group. Conflicts of interest will surely arise and, with such a manager as boss, the other groups will certainly suffer. A common cause for conflict is access to equipment, with the boss's group being favored. This situation gets worse when one of the groups is doing R&D in the same or similar fields as the manager's special group. This is an unhealthy circumstance, probably leading to a messy affair, loss of several senior people, and loss of confidence in top management. A major concern should be good management, not kudos for the manager for hands-on work, and the success of all groups under a manager should be a major source of that person's satisfaction.

To get things done through others, managers must influence others.

To do this, *good managers are sensitive to the needs and attitudes of people.* They have a sixth sense for the needs of people and for group problems. There are some managers who become so concerned with the technical problems of a project that their antennae are not out for possible personnel problems.

Good managers also communicate well horizontally and vertically and have a talent for listening with empathy. I've seen people listen intently and then respond on another topic. Their thoughts were elsewhere; they merely put on an excellent facade portraying interest. Sooner or later, they lose their audiences and the cooperation of these audiences.

Good managers are creative; that is, they have the ability to analyze, synthesize, and imagine. Creativity is an essential part of a manager's makeup. With a creative bent, a manager can participate with colleagues in significant discussions and can make exciting decisions.

And, of course, *good managers have an inherent intellectual capability* and an ability to comprehend technical situations. Criteria for selecting managers include above-average technical expertise and exhibited interpersonal skills.

Influential managers also have an entrepreneurial bent, in that they constantly look for new opportunities for their groups or organizations. This entrepreneurial bent means taking chances, for only with risk comes reward.

Good managers have personalities that cause people to listen to them; good managers do not turn people off by an abrasive, arrogant, dull, or otherwise negative or offensive personal projection. Ailes[28] says that it is important to be liked. Of the several qualities of likable people, I think optimism and the acceptance of others at face value are very important. These lead to a climate that seems to make people more productive. Rude, tough leaders can run profitable enterprises, but sooner or later their style of management will alienate subordinates, leading to a climate that makes people less productive. I remember one such boss who took great joy in putting down a professional in front of others. His power to do this seemed to inflate his ego, but he never won any points by such actions. You've also probably heard some managers say, "I'm not here to win a popularity contest." True, yet it won't hurt to be liked. When you lose support of subordinates, you lose leadership. While being liked is important, good managers also strive to be respected by hard work and solid performances.

Accent the positive!

Personality individualizes the manager; it is one of the person's identity characteristics. Other such characteristics include grooming, appearance, quality of voice and expression, and personal hygiene.

The identity stamp sets a person apart from others; it can be improved but only if the person recognizes the need and then wants to improve.

Let them know you for your winning personality.

Managers must also be credible[29] in order to execute authority effectively. Their staffs must see them as competent, honest, strong in character, consistent in action, fair, sensitive to surroundings, mentally flexible, and capable of delegation. Managers also enhance their credibility when they support their staffs while simultaneously setting high personal standards with the expectation of having them emulated. It is incredible that some managers attempt to outdo their staff in some mental exercise rather than working with them to produce a better answer to a problem. Or, how about a manager who lectures a nonprofessional for an hour or more when a 5-minute discussion might have accomplished something positive? Incredible acts make the manager incredible not only to the person involved but, via the underground, to everyone.

Managers must be credible to be effective.

Skills of the effective manager

Managers are leaders having authority, although good leaders have the ability to influence people even without authority. We want managers who are good leaders, and this depends on their having a reasonable balance of technical, administrative, interpersonal, and leadership skills:[17,20,30]

- *Technical:* A broad technical background supports creative ability and the ability to judge new ideas. Managers' technical knowledge may not be as comprehensive or as current as that of their staffs, but they understand technical discussions and can ask probing questions. They can dissect an issue and get at fundamental cause and effect relationships, can conceive or appreciate new uses for new technology, and are able to participate in brainstorming sessions.
- *Administrative:* Required is knowledge of budgeting and cost control methods, personnel relations, organizational and planning methods; an ability to make decisions in a timely manner; and patience with paperwork.
- *Interpersonal:* Required is the ability to motivate, to mediate, to influence, to communicate effectively, and to listen with empathy. Badawy[30] has shown that organizations fail because managers fail, managers fail because they perform poorly, and the major cause of

poor managerial performance is poor interpersonal skills. Hymowitz[31] lists five reasons why managers fail including their inability to get along with others.

- *Leadership:* Leadership skills combine the above skills, especially technical and interpersonal skills, and give rise to an ability to obtain and retain highly capable staff, to communicate effectively, and to motivate and influence others.

In each of the above categories, there is one attribute that, I believe, sets one manager apart from others. Under technical skills, I was always impressed by a superior's ability to take charge of a meeting and to identify the important issues by exerting constant pressure to get at fundamental cause and effect relationships. This directs professionals toward the important issues and away from a general technical discussion.

Among administrative skills, I select the ability to make timely decisions. There is the story of the director of research who always took a long walk when faced with a tough decision. If a walk is taken to clear the mind to better formulate a decision, OK; if it is to delay a decision, NG. Whatever it takes, decisions must be made in a timely fashion if managers are to remain credible.

Decisions, decisions, decisions.

Among interpersonal skills, the ability to communicate effectively is a wondrous talent. Unfortunately, this ability can be used for personal aggrandizement, one-upmanship, or "doing in" the opposition. Used positively, it can be an exceptional talent.

Leadership is made up of many facets, but the ability to influence others marks the leader, whether line or staff.

Approaches to managing people

Good managers must manage time, money, and resources. Most importantly, they must manage people. In the next few pages, I introduce some management theories and then conclude with how managers should manage people.

Management Theories.[22,24,25,32] The first well-developed, formal theories of management were advanced in the late 1800s and early 1900s. The *mechanistic classical management theory* had its roots in the belief that workers had no valuable input and needed the continual guidance of management. It was a very impersonal approach toward making the worker (and, subsequently, the organization), more efficient. In the 1930s, the mechanistic classical management theory gave way to the *behavioral school* of management, and employee behavior became the focus of ways to increase productivity. The works of McGregor and Hertzberg are particularly important in understanding today's behavioral theories.

Management theories are reference points for one's personal actions.

McGregor's theory X and theory Y[33] have substantial differences. Under *theory X*, management gives firm direction to, and has close control over, personnel. People are told what to do and how to do it, and are monitored during work so that rewards or punishment may be accorded. This approach assumes that the average person dislikes work, prefers direction, avoids responsibility, is not very ambitious, is unconcerned about the company, and is motivated primarily by a need for security.

Under *theory Y*, management's concern is with relationships, with the working environment, and with the fostering of initiative and self-direction. This approach assumes that people like to work, exercise self-direction and self-control, seek to achieve goals in order to gain satisfaction from personal achievement and tangible and intangible rewards from management, accept and seek responsibility, have a creative capacity, and only partially realize their potential.

What about a continuous management spectrum from X to Y?

Hertzberg's motivation-hygiene theory[34] proposes that satisfiers and dissatisfiers exist. In this theory, satisfiers that determine job satisfaction are such things as achievement, recognition, the work itself, responsibility, and advancement. Dissatisfiers include company policy and administration, supervision, salary, interpersonal relationships, and working conditions. Satisfiers are intrinsic to job content and lead to satisfaction because of the need for growth or self-fulfillment; dissatisfiers are extrinsic to the environment and lead to dissatisfaction because of the need to avoid unpleasantness.

From the 1960s on, various people have proposed management approaches that are contingent on each circumstance. *Contingency theories* integrate previous theories. According to Griffin,[22] the path-goal theory of Evans and House (advanced in the 1970s) suggests that leaders will clarify the paths that an individual must take to achieve goals and reap rewards and assumes that the leaders can change their approach to meet the demands of varying situations.

Pay particular attention to the various contingency theories.

About the same period, Morse and Lorsch[35] offered their contingency theory based on a necessary fit among tasks, organizations, and people. People vary in needs and motives but seek a sense of competence. They find that when a fit is obtained, both effective unit performance and a higher sense of competence result. Also about the same time, Allen[36] suggested theory M as an alternative to theories X and Y. In theory M, management actions vary as the needs of people and situations vary. More recently, Stanton[37] has expressed the need for increasing productivity in the face of a lessening will of employees to work. He proposes reality-centered management, which takes into account the various types of employees and recognizes their differing abilities and needs. Because of the many variables in this approach, Stanton believes that managers must be flexible, using a participative approach with some people and a more direct approach with others. He also describes the need for effective recruiting, placement, training, and appraisal; for effective supervision; and for an equitable reward and compensation system.

Knowledge of management theories and their advantages and disadvantages places a manager in a superior position to define and (or) to question his or her own approach. The above theories are not the only ones advanced during the last decade, and the "References" and "Suggested Readings" (cited at the end of this chapter) provide the curious reader with further information.

Autocratic versus democratic management.[35–37] Management theories cover the spectrum from very autocratic to very democratic approaches toward managing people. There are situations where workers want and need direction, such as in piecework, assembly, and other routine tasks. In R&D, participatory leadership was shown over three decades ago to be superior to laissez-faire or directive leadership.[38] The benefits of participative management are well documented;[39] however, participation does not mean that an organization uses theory Y in its extreme. You can have varying degrees of participation—from cooperation on the job to participation in managerial decisions. You want participation in the managerial decision *process* but not necessarily in the making of the final decision. You also want participation in order to take advantage of what everyone has to offer, not primarily to satisfy everyone. Obtaining satisfied employees is a secondary, although gratifying, result.

Managers can be autocratic, democratic, or cover the spectrum.

The head of R&D will find it appropriate to accept some of the suggestions of managers as correct decisions. In other situations, he or she will find it best to make the decision. The professional staff should make ongoing decisions pertaining to their assigned projects. They should participate in their area's decision process, but the managers should be the decision makers. At every hierarchical level, participation should be occurring, but staff and area managers must recognize that their manager is the boss. No manager is "one of the gang." During my visits to Japanese firms, I saw participation in action, but my Japanese colleagues and I still knew who was the boss.

I suggest that managers must be flexible in their approach to managing. People are different and must be approached as individuals who can change behavior over time and as situations change. For example, a technician may want and need direction, whereas a professional may require only delineation of responsibility and authority. Thus, motivation methods must change to meet the varying needs of people. A manager's concern for the welfare of his or her staff is second only to his or her concern about the work the staff is producing and whether the work supports the goals of the group's strategic plan (described in detail in Chap. 3). This being true, the contingency approach described above fits an R&D manager's needs.

A flexible manager means a responsive manager.

The management theory you select and pursue becomes a significant aspect of your management style and dictates how you treat your staff and how your staff responds to you and the company. Your characteristics and the skills you have acquired must support your ability to influence and motivate those role players who make up the manager's environment.

A reasonable balance of skills and characteristics is needed.

An R&D Manager's Environment

A manager's environment entails interaction with several entities, organizations, and persons. It has two aspects: that environment external to the company and that internal to the company. Some of of these interactions are listed below:

An R&D manager's environment is more global than just the R&D group.

- External environment

The local, national, and international technical communities (such as universities and associations)
The local and national political and legal communities (such as regulatory bodies)
The market community (such as competitors, suppliers, and customers)

- Internal environment

The R&D staff, including managers, professionals, technicians, and support persons
The boss, including an R&D manager's superior, the operations' vice presidents, and their staffs
The support groups, including accounting, personnel, and safety
The managers in downstream groups, including marketing, engineering, and manufacturing, and the division presidents responsible for these groups

You will recognize that all the role players in the innovation process have just been listed under the R&D manager's internal environment. It would seem logical that an R&D manager should have ongoing associations with everyone listed above. However, a not atypical relationship is that shown in Fig. 2.1*a*, wherein a linear relationship exists between boss, R&D manager, and the R&D staff. Such a linear relationship may be augmented by occasional meetings with downstream groups (represented by the dotted line), but this arrangement will never succeed in getting R&D results into the marketplace. This may be management of invention but not management of innovation. Invention, or the production of something new, is necessary but the invention put to use in the marketplace, or innovation, is where profitability lies. We must examine further the R&D manager's internal environment listed above and expand the figure to that showing continual working relationships that can lead to success. In Fig. 2.1*b*, we have a network of managers who play roles in the innovation process and with whom the R&D manager must bear the primary burden of establishing and promoting collaboration. Indeed, we can liken this to the Japanese quality circle, or to use more exact nomenclature, I call it "innovation's management circle." This emphasizes the need for intercommunication and collaboration among all managers in the innovation process. This is not a casual relationship; it is meant to be dynamic and continuing, one intended to make the innovation process work successfully.

In such relationships, R&D managers must be open and straightforward. Don't scheme, don't play games, and don't practice one-up-

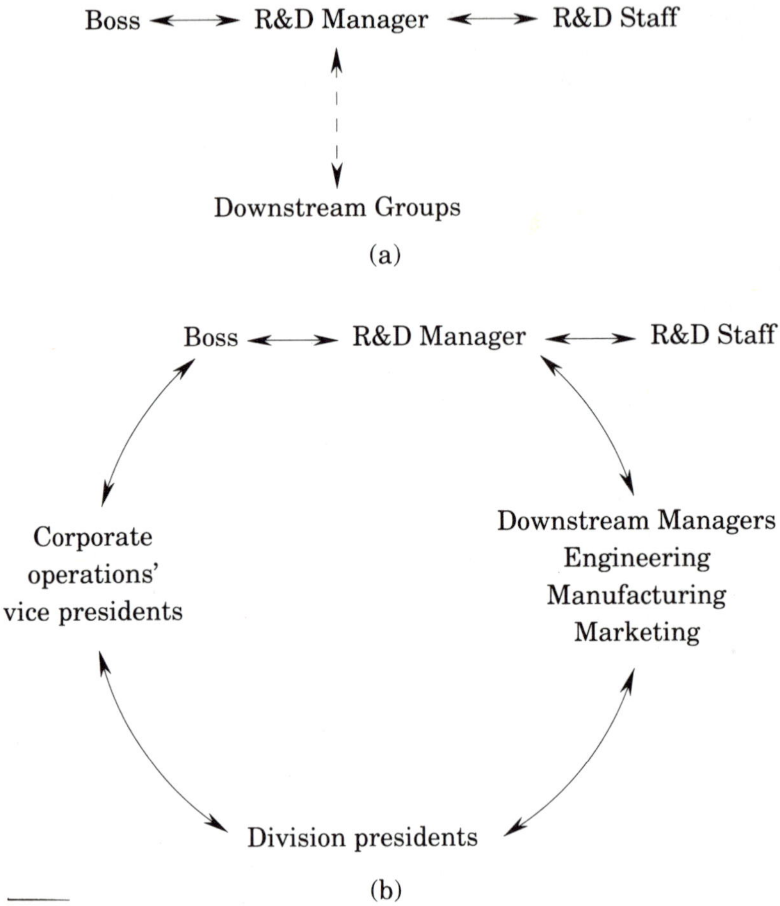

Figure 2.1 A comparison of static and dynamic innovation management: (*a*) a linear management relationship, (*b*) innovation management circle.

manship with others in this network, or you'll sooner or later have no open channels of communication. The consequence will be quiet ostracism and nothing from your group will get commercialized.

In the ideal situation, every operational group would do the planning and implement its plans, and R&D would obtain the cooperation and collaboration of these groups. There would be interactive communication and participation throughout the management circle, and the goals of management would be carried out routinely and successfully. The totality of the corporate energies would be applied to innovation just as it is to financial aspects. Of course, there is no such utopia; we can only strive to achieve it. R&D managers must take the initiative to obtain collaboration of these role players.

Utopia? Was ist das?

Characteristics of Those We Want to Influence

Although the characteristics of managers vary in degree, they must include those discussed in the section on management style: the need to manage (the desire to supervise and influence others), a sensitivity to the needs and attitudes of people (to influence them, you must know them), the ability to communicate (to convey and to listen), creativity (the ability to participate with colleagues and to make decisions in a timely manner), an inherent intellectual capability (evidenced in above average technical and interpersonal skills), entrepreneurial bent (the ability to seek new opportunities and stimulate the staff to do the same), a positive personality (a personal projection that attracts, detracts, an identity characteristic), other identity characteristics (appearance, voice, and the like—all subject to praise or damnation), and credibility (to be leaders, managers must be believable).

The boss and the corporate staff

In a corporate R&D group, your boss may be the president and CEO or his or her designate, perhaps an executive vice president. Activities are less on technical aspects and more on financial and planning activities. The president also leads in the establishment of policies and procedures and in the definition of corporate goals. Among these goals are those specific to R&D. These are obtained by or communicated to the vice president of R&D, who in turn communicates them to subordinate managers and staff. If goals are not delineated, you must devise a set of goals and pass them by your superior for editing and approval.

The president-CEO has a balance of the above characteristics along with certain attitudes and beliefs that directly impact R&D. The short-term versus long-term attitudes of a president-CEO establish the character of the R&D group. Early background as a scientist, engineer, or nontechnical person may influence the degree of patience with which R&D is treated and affect what is believed to be realistic schedules of accomplishment. These, in turn, may or may not be realistic to the R&D group but will surely establish the type of R&D that is acceptable or possible.

Vice presidents of operations are managers who link the CEO to operations. Each may have responsibility for several divisions of common interest. They interpret corporate goals for the divisions and are concerned with the divisions' short- and long-term welfare. Whereas they monitor the bottom line, they must also monitor external threats and opportunities. Such opportunities include acquisition, new ventures, and new product concepts.

The style of management that your boss exhibits is most important to the climate of the company and the R&D group. How does your boss go about securing a productive relationship with subordinates? What is his or her opinion of R&D people? Does a big ego require a flaunting of power or does your boss simply lead? Does your boss rule with an iron fist or inspire through the perspiration of example or through the influence of personality? Is there a concern for people along with high performance expectations?

How you communicate with and attempt to influence your boss is dictated by *his or her* characteristics, skills, and approach to management.

Managers in downstream groups

The division presidents are managers who have much the same activities as the vice presidents of operations, but they are more intimately involved with the day-to-day struggles for profits. They are more focused on the peculiar interests of their divisions. They have closer relationships with their engineering, manufacturing, and marketing managers and share their concerns and visions. These division presidents have to be hard nosed about letting their people start a new venture or R&D program and must see that it offers reasonable probability of new profits, growth, and leadership. Thus, they have the obvious responsibility to the operating vice presidents to see that plans are implemented efficiently and that projections are reached. By both formal training and industrial experience, the professionals and managers of the engineering, manufacturing, and marketing groups have attitudes and goals different from those of the R&D staff. For example, they tend to be more consumer oriented or market-driven, while R&D groups tend to be more knowledge oriented or technology-driven. These persons, being closer to the consumer, may even depreciate the R&D group as being in an ivory tower, not understanding the needs and urgencies of real life, and certainly not understanding what makes a practical product. This conflict of interest needs bridging by leaders of the organizations involved.

It takes two to tango.

As an R&D manager, you want coupling between your R&D group and the downstream groups. The goal of all coupling is to bring the greatest amount of intelligence to bear on the understanding and solution of problems so that participants obtain maximum satisfaction, practical solutions emerge, and the continuity of the process is assured. Lack of coupling means that not all company capabilities are being tapped and there is increased chance of failure. Ansoff and Stewart[40] suggest that one characteristic of successful high-technology businesses is the degree of downstream coupling; that is, the success of a company's product introduction depends on communication and cooperation between the R&D function and the engineering, manufacturing, and marketing functions, which are downstream toward the customer. The engineering, manufacturing, and marketing managers are where the action exists that will ultimately reflect on the company welfare. Yet bonuses depend on profits, so they are sensitive to setbacks whether in the marketplace or in an R&D program. They are prone to push the known over the unknown or a product improvement over a new product. So a balanced portfolio of short- and long-term projects requires an enlightened management that encourages and provides rewards for balanced portfolios that yield success.

The staff

Although the R&D staff are not part of the network of managers, their results are passed from manager to manager through the innovation process. Hence, R&D managers must have their staff's complete support for the group's mission and goals.

The R&D manager works with technicians, professionals, support personnel, and other managers. I'll never depreciate the value of an outstanding technician or secretary, but it must be said that the innovation process depends in the main on a creative staff of professionals. That means we ought to know more about the typical professional.

Professionals have spent a minimum of 4 years in college, and many have spent as many as 10 years in undergraduate and postgraduate training. With such an investment, it is understandable that they exhibit a strong need for achievement and recognition. Furthermore, the transition from university to industrial life can be a disturbing adjustment. University research done at any time of the day or night suddenly becomes an 8-5 (or -6 or -8) job, and management must channel the professionals' efforts so that they meet organizational as well as personal goals. Of course, there is always the exception; common sense must rule. Remember the story "He Marches to a Different Drummer"? Rather than acting independently (especially as to time and research content), the professional new to industry is now one cog in a larger investigative team. He or she can no longer pick research on the sole basis of personal interest but must do research within certain prescribed limits of the group's mission and goals. This may lead to some strain in initial relationships with managers. Managers may be regarded as technical persons gone wrong, probably prostituting themselves and their profession for the dollar, or they may be seen as glorified paper shufflers who are a necessary evil to support the professional's endeavors. It is also true that some managers never remember that they were once new professionals. A major problem arises, of course, when the professional never gets the rough edges worn off and, although never reaching an understanding of industrial management, never leaves the industrial life.

Early training affects outlook and beliefs.

Managers can be OK people.

A company usually attracts professionals because of the company image, the mission of the specific group they might join, other staff members' personalities or reputations, the type of work they might do, the freedom they might have to create within some mission limits, the degree of self-management allowed, the pay and incentive programs, and the geographic location.

Von Glinow,[41] through a comprehensive review of the literature, concluded that a professional orientation rested on:

- Obvious *expertise*—the result of specialized training
- A code of *ethics*—the rendering of a service without personal concern for reward
- *Maintenance of standards* among colleagues—control by discipline group rather than governmental or hierarchical control, since professionals claim adequate expertise to police their specialty
- A desire for *autonomy*—the freedom to work on projects professionals deem important and to work in their own way (given the limits of the group's mission)
- An ability to make a *commitment*—the commitment to a technical specialty and to a career
- A professional *identity*—the need to identify with peers

How Do You Get Collaboration?

Obtaining the collaboration of others depends heavily on four personal attributes:

- You must have characteristics that facilitate discussion.

- You must have skills that command respect.
- You must have an approach to management that shows respect for the needs of others while expecting high standards of the work ethic.
- You must know your audience.

To obtain collaboration, others must see you as being an intellectual person, as having a good personality, and as being credible. No one wants to waste time talking to a dullard or to an abrasive or ill-kempt person or to one whose acts seem incredible. The grapevine lets everyone know what type of person you are.

To get your message across, you must have the essential skills of a good manager and, especially, adequate technical and communications skills. These are obvious from the need to convey a concise technical message.

Your management approach affects others' decisions on whether they will or will not collaborate. Professionals or managers under your jurisdiction want to make intellectual contributions, not just physical contributions. They want to participate in the decision-making process. And they want their authority and responsibility spelled out. You should propose these matters for consideration and not wait for them to broach such thoughts—or worse, have them go unsaid and undecided.

Be a responsive leader.

Finally, you must know the attitudes, beliefs, eccentricities, needs, problems, and other characteristics of your audience. The greater appreciation and understanding you have of their circumstance, the easier it is to have two-way communication.

Collaboration with your boss is extremely important. You must seek real agreement on mission and goals and on how your R&D group fits into the corporate structure and plans. With a sincere commitment and mutual understanding of the role of R&D, the following will accrue:

1. Downstream groups will be caught up in the climate and will be more amenable to consider ideas from R&D.
2. There will be an integration of R&D into the "family," thereby moving the company forward. (Beloved is the CEO who finds this important.)
3. Such an environment will enhance the R&D leader's confidence in

his or her group's ability to produce. (If you are always on trial, you will never produce.)

Keep operations vice presidents and division presidents informed of progress in the R&D group. Show the relevance of the R&D work to their interests. But it's a two-way street; give in order to get. Ascertain their hopes for one to five years out and discuss how they expect to get there. Can you propose how R&D might assist? Follow up such meetings with summaries, provocative suggestions, and information on new technologies. Get them to depend on you and your group for such information.

Take a promising new idea to the downstream managers of engineering and marketing. If favorable reaction is received, then you can describe it to the operations vice presidents and division presidents. But don't close the door on good ideas by doing this vice versa. I'll have more on coupling with managers in downstream groups in the "Plan Ahead for Technology Transfers" section in Chap. 4.

As discussed in the previous section, the narrowness and the individualistic nature of training received by professionals runs contrary to the company's need for a multidisciplinary approach to solve complex industrial problems. Their needs will differ depending on their individualities, the work situation, and their experience. The consequences of the professional's self-determination (autonomy) include a strong need for participation in goal setting and in the decision-making process. Hence, to get collaboration of your professionals, you must know them. There are several additional actions managers should consider in dealings with professionals:

- Provide a clear understanding of corporate goals, of the mission and goals of the group, and of where the professional's work fits in (obtained from the strategic plan described in the next chapter).
- Support your staffs, and provide the environment conducive to working effectively to their own and the company's profit.
- Let your professionals know that you care about what happens to them as well as how their work is progressing.
- Show the benefits of working with others and using others' knowledge.
- Convey enthusiasm and optimism for their activities.
- Open doors for their accomplishment.
- Provide feedback from the decision-making process.
- Provide for continuing education and retreading.

- Provide constructive advice without deflating egos.
- Acknowledge a job well done, and reward performance.

You must also seek the collaboration of support groups. They are not within the management circle, but they can be of great assistance to your efforts. Persons in accounting and personnel are professionals specially trained for their specific jobs. Since they are service groups, they receive a lot of flack. They are likely to be sensitive to questions and particularly conscious of hierarchical protocol. If answers to your questions go up to their top management and then across to yours, they may or may not filter down to you. Make certain that answers are handled according to your desired way when posing the questions.

Summary

- Management style starts at the top and becomes part of the corporate image and culture.
- An individual's management style is a composite of characteristics, skills, and management approach and is reflected in the results of the person's activities.
- Good managers have the following characteristics:

  They need to manage.
  They understand the needs and attitudes of people.
  They are good communicators.
  They are creative.
  They have an inherent intellectual capability.
  They have an entrepreneurial bent.
  Their personality is such that people listen.
  They are credible.
- Good managers have a reasonable balance of technical, administrative, interpersonal, and leadership skills.
- Good managers have a flexible approach to managing people contingent on their needs and the circumstances.
- Managers must influence those within their management network to obtain the results desired, and this is done through management style and a knowledge of their audiences.
- A manager's environment comprises external and internal associations which must be nurtured.
- The concept of innovation's management circle emphasizes the internal groups necessary to a manager's success. The R&D manager

must take the initiative for actions pertaining to an innovation's progress through the six-phase process.

- Success in innovation depends on mutually supportive managers working together.

Cases and Incidents

The Starter and the Manager Who Can't Say No

Joe is a metallurgist in Larry's area and has a creative flair. He's a loner, likes to work after hours, and doesn't trust people from operations to carry on with his developments. Larry gets him started on applying a new coating to a gear part which would ostensibly reduce wear, lengthen life, reduce resistance in its housing, and be less costly than the current product. This is not, however, a high-margin item, and the current product sells. So, Larry has Joe determine what needs to be done and how long it might take him.

Joe gets the first product out fast. He jerry-rigs a setup that allows him to apply the new coating, but the part warps a bit. The coating looks promising but needs extensive grinding to get the right tolerances and a good finish. Joe is optimistic that he can overcome the disadvantages and sets about changing his setup to allow application with less total heat pickup by the part. OK, but the agreed-upon time is up; Joe avoids everyone. Larry finally corners him and asks about results.

Joe says he doesn't like the appearance of the coating and needs more time. When Larry tells him he must have some test pieces for a division test, Joe promises them in two weeks. For sure.

Larry reluctantly goes along with this. A month goes by and still no samples. Joe says that the part ought to be redesigned and that the setup needs to be changed so that there is better control of oxidation.

Well, Larry can't say no; he understands the technical reasoning so he lets Joe proceed. However, all this gets to Helms, and after much deliberation, he gives Larry the word that this is Joe's last extension—produce or else.

Helms and Watts talk about the situation. Helms wants to know what Watts's ideas are. Watts says his experience indicates that with people who can get something started, who have some good ideas but who never get a project completed, you usually have to get the project into other hands or get someone to work with them who is a known "finisher." If you can't do one or the other of these, you have no choice but to cancel the effort.

Helms asks Watts what he thinks of Larry as a manager. Watts comments that Larry is creative and personable and that his staff likes to work for him. However, he is too passive and could enhance productivity by being firm in his expectations of accomplishment. Watts says Larry could get greater respect from his staff by being firm (which would probably come as a surprise to Larry!).

Helms does not see any parallel between Larry and himself.

Questions

- Is Joe valuable? If so, how do you make him productive?
- Would Joe accept working with a "finisher"? What could you do to get him to go along with such help?
- Would a discussion with Joe help? By whom?
- How about a discussion with Larry about managing people? About matching people's strengths and weaknesses? And about setting goals with his staff that he expects them to keep?

For Want of a Decision

Harry's visitor is an inventor well known to him. The inventor looks at the ceiling a lot while he talks, is carelessly dressed, is only moderately clean, and has a long and sometimes confusing story about his latest baby. The baby turns out to be an analytical device to measure ammonia in trace concentrations. Harry (senior electronics engineer in Larry's area) sees no holes in the concept and brings in Helms for confirmation. Helms considers the concept valid and phones long-distance to Jones, the chief engineer of XYZ's instrument division, who is immediately interested.

Jones flies out, and another meeting is held with the inventor. All goes well, everyone is enthusiastic, and Jones prevails on the inventor to fly back with him to make a presentation to the division president. The meeting with the president also goes well, but the president does not quite trust the inventor, because of his appearance and random speech. The inventor leaves, saying that he'd like to go with this company but hopes a decision can be made soon because he has others also interested.

The president wants a consultant to look over the data and give a review, so Jones gets this done in the next two weeks. Results are positive, and Jones begs the president not to delay his decision: "We've got to move on this one or we lose it."

The president, not to be pushed, says, "Oh, these guys always say they have someone else in the wings to get us to leap before we've examined their 'baby' thoroughly."

Jones doesn't agree and presses again for a quick decision. Finally, the president says he'll sleep on it and get back to Jones. With that, he takes a long weekend. On his return, he tells Jones to go ahead. Jones says that it's too late, that Harry called Friday, saying the inventor made a deal with the competition.

Helms tells Watts the story and says they lost a good deal because that cluck couldn't or wouldn't give Jones a decision.

Watts surprises Helms by saying that Harry and Jones also made a big mistake. Helms and Harry should have gone over every detail with Jones, and then, if the deal was positive, Harry and Jones should have presented the story to the president asking for an immediate decision. The inventor should have been left out of the meeting. So the loss wasn't entirely due to want of a decision; it was also due to letting the president meet the inventor!

Watts also says, "As to decision makers, there are some who defer in the face of mountains of justification, there are those who go through requests with a red pencil faster than the pencil can keep up, and in between we have managers who consider data carefully and make timely decisions, and we thank God for them."

Questions

- Isn't knowing your audience and the participants important to winning?
- If you have all the data, are there still reasons for not making a decision?
- Should the president procrastinate in the face of recommendations from a senior member?
- Did the president lose more than this opportunity?

References

17. D. B. Miller, *Managing Professionals in Research and Development,* Jossey-Bass, San Francisco, 1986.
18. M. A. Von Glinow, "Reward Strategies for Attracting, Evaluating, and Retaining Professionals," *Human Resource Management,* **24**(2): 191–206 (1985).
19. M. Maccoby and H. Brooks, "Why Management Style Has to Change," *Res. Mgmt.,* March–April, 1986, pp. 44–45.
20. R. Brown, *The Practical Managers' Guide to Excellence in Management,* AMACOM, New York, 1979.
21. F. C. Dyer, *Executives' Guide to Handling People,* Prentice-Hall, Englewood Cliffs, N.J., 1959.
22. R. W. Griffin, *Management,* Houghton Mifflin, Boston, 1984.
23. D. C. McClelland and D. H. Burnham, "Good Guys Make Bum Bosses," *Psych. Today,* December 1975, pp. 69–70.
24. R. E. Miles, *Theories of Management,* McGraw-Hill, New York, 1979.
25. M. H. Mescon, M. Albert, and F. Khedoun, *Management: Industrial and Organizational Effectiveness,* Harper & Row, New York, 1981.
26. E. Raudsepp, "Profile of the Effective Manager," *Chem. Eng.,* Mar. 27, 1978, pp. 141–144.
27. D. C. McClelland and D. H. Burnham, "Power Is the Great Motivator," *Harvard Bus. Rev.,* March–April 1976, pp. 100–110.
28. R. Ailes with J. Kraushar, "The Importance of Being Liked," *Reader's Digest,* May 1988, pp. 133–135.
29. "The Manager and Credibility," *The Dresser Manager,* Dresser Industries, Dallas, Tex., April 1982.
30. M. K. Badawy, "Why Managers Fail," *Res. Mgmt.,* May–June 1983, pp. 26–31.

31. C. Hymowitz, "Five Reasons Why Managers Fail," *Wall Street J.*, May 2, 1988, p. 25.
32. D. Hellriegel and J. W. Slocum, Jr., *Management,* 3d ed., Addison-Wesley, Philippines, 1982.
33. D. McGregor, *The Human Side of Enterprise,* McGraw-Hill, New York, 1960.
34. F. Hertzberg, *Work and the Nature of Man,* World, New York, 1966.
35. J. J. Morse, and J. W. Lorsch, "Beyond Theory Y," *Harvard Bus. Rev.*, May–June 1970, pp. 61–68.
36. L. A. Allen, "M for Management: Theory Y Updated," *Personnel J.*, December 1973, pp. 1061–1067.
37. E. S. Stanton, *Reality-Centered People Management,* AMACOM, New York, 1982.
38. H. Baumgartel, "Leadership Style as a Variable in Research Administration," *Adm. Sci. Q.*, December 1957, pp. 344–360.
39. W. M. Jeuchter, "The Pros and Cons of Participative Management," *Mgmt. Rev.*, September 1982, pp. 44–48.
40. H. I. Ansoff and J. M. Stewart, "Strategies for a Technology-Based Business," *Harvard Bus. Rev.*, November–December 1967, pp. 71–83.
41. M. A. Von Glinow, "Incentives for Controlling the Performance of High Technology and Professional Employees," *IEEE Trans. on Systems, Man, and Cybernetics,* SMC-13(1): 70–74 (1983).

Suggested Readings

Badawy, M. K.: "How to Succeed as a Manager," pt. 1: "The Manager's Role," *Mach. Design,* March 1981, pp. 76–79; pt. 2: "The Causes and Prevention of Failure," *Mach. Design,* April 1981, pp. 217–221; pt. 3: "Why Switching from Engineer to Manager Is Difficult," *Mach. Design,* May 1981, pp. 91–95; pt. 4: "Taking Charge of Your Career," *Mach. Design,* May 1981, pp. 59–63.

Donnely, J. F.: "Participative Management at Work," *Harvard Bus. Rev.,* January–February 1977, p. 117.

Drucker, P.: "The Best R&D is Business Driven," *Wall Street J.,* February 10, 1988.

Greene, C. N.: "The Satisfaction-Performance Controversy," *Bus. Horizons,* October 1972, pp. 31–41.

Hayes, R. H., and W. S. Abernathy: "Managing Our Way to Economic Decline," *Harvard Bus. Rev.,* July–August 1980, pp. 67–77.

Hull, D. L., J. J. Bosley, and G. G. Udell: "Renewing the Hunt for the Heffalump: Identifying Potential Entrepreneurs by Personality Characteristics," *J. Small Bus. Mgmt.,* January 1980, pp. 11–18.

Ouchi, Wm.: "Theory Z Corporations: Straddling U.S. and Japanese Molds," *Ind. Week,* May 4, 1981, pp. 49–52.

———: "Ouchi's Theory Z... A Meeting of East and West," *Mgmt. Rev.,* May 1981, pp. 8–16.

Stumpe, W. R.: "Entrepreneurship in R&D—A State of Mind," *Res. Mgmt.,* January 1982, pp. 13–16.

Part

2

Technical Management Activities

There are many definitions of management in the literature based on a manager's activities. These can vary substantially depending on an author's emphasis. We want a good working definition for our technical environment, and that which I feel best fits is "Technical management is the process of planning, implementing, leading, and controlling a group's resources to achieve goals in an effective manner." According to this definition, a responsible (and responsive) manager (and I mean managers at all levels) does the following:

Plans. *Sets goals, formulates strategies, oversees project selection and portfolio formulation, and looks for new opportunities. (See Chap. 3.)*

Implements. *Assigns major responsibilities, organizes people into teams, obtains the necessary resources, makes certain that work progresses well and in the best interests of the company, and looks for new ways to increase performance. (See Chap. 4.)*

Leads. *Obtains proper staff by recruiting, training, and appraising; communicates laterally and vertically while disseminating information; motivates by inspiring staff, encouraging performance, acting as an entrepreneur, and promoting and supporting the staff and their ideas; and looks for new ways to use staff. (See Chap. 5.)*

Controls. *Monitors progress, stimulates change, allocates resources, oversees budgeting and cost control; and looks for new ways to increase efficiency. (See Chap. 6.)*

Part

2

Chapter

3

Planning Activities

Questions answered in Chapter 3

- Why is strategic planning vital to the success of an R&D group?
- How does the R&D plan fit with the corporate plan?
- Who makes up the planning team?
- What are the six stages of the planning process?
- How is collaboration among divisions, departments, and groups promoted by the planning process?

Planning is the first of the four managerial activities discussed and is primary to initiating and targeting the innovation process. *Good* planning commands the participation of all role players involved in the innovation process (innovation's management circle).

In the following section, I review corporate planning, which involves corporatewide activity that culminates in a plan of action for the corporation. One part of this plan is the corporate R&D plan. Because R&D is so important to the continuing welfare of the corporation, a separate R&D plan should integrate all R&D within the corporation. Unfortunately, this is not the case in many companies, since it involves additional overhead time and raises the specter of internal conflict. As a result, operational R&D plans are issued with the division plans, and the corporate R&D plan is just another corporate departmental plan.

Corporatewide plans must fit into a mosaic that meets company needs.

In the strategic planning section, I discuss in detail technical planning, taking into account the need for collaboration of all relevant groups. Although I describe a formal planning method that has a beginning and an end, good managers and professionals plan continually. Even in daily activities, they think about an action, alternatives to the action, and all possible consequences before acting.

Any action warrants some prior planning effort.

An Overview of Corporate Planning

Companies vary in the degree of sophistication with which they accomplish planning. In Gluck, Kaufman, and Walleck's[42] four-stage evolutionary model of company planning systems (see also Hoch):[43]

- Stage I companies have basic financial planning systems in which everything is reduced to a financial problem and the value standard is meeting the budget.
- Stage II companies extend basic financial planning by means of long-range forecasts.
- In stage III companies, planners try to understand the marketplace phenomena forcing change, look for opportunities that may lead to a more attractive portfolio, and present alternative strategies to top management.
- In stage IV companies, management is involved in strategic planning that stimulates entrepreneurial thinking and promotes managers' commitment to the company's strategic plan.

I like this four-stage model because it permits an easy segregation of companies and an evaluation of where companies are in the spectrum of planning sophistication. Unfortunately, some companies think they are at stage IV, when in reality they are at an earlier stage. Corporate management should give some thought to identifying the degree of sophistication they have reached, what their planning exercises are yielding and why, and at what stage the company should really be. Any company that derives a plan but does not follow that plan (or constantly changes the plan without any sound basis) not only is wasting time and money but also is frustrating and discouraging its employees—a real turnoff! Of course, environmental changes may force a change in plans, but a company that plans and backs the plan with a corporatewide commitment is enhancing its chances for success.

Strategic planning combines corporatewide collaboration, entrepreneurial thinking, and commitment.

In general, planning includes two planning periods: first, the development of the strategic plan, which defines and communicates longer-term business directions and, second, the development of the operating plan, which specifically identifies the tasks or projects (tactics) that will implement the strategic plan.

The strategic plan

In planning groups, the popular phrase for some time has been "strategic management,"[44] which implies consideration of both the external environment and the internal development of capabilities to carry out the strategies developed. The main difference between previous long-range planning methods and current strategic planning methods is in the development of alternatives based on continuing forecasts of the future. In strategic planning, the planners develop and analyze alternatives for every facet of a business area, such analyses leading eventually to strategy choices. These alternatives take into account possible futures, as perhaps developed in scenarios made up of hypothetical future events, from which impacts can be derived and strategies proposed. Thus, by looking into the future, the planners are in a better position to make decisions that will affect the future.

Strategic planning considers alternative strategies.

Today's decisions affect the future, so attempt to make good ones.

All corporate departments, operating divisions, and companies must develop plans. In each division, R&D, engineering, manufacturing, marketing, sales, and all support groups participate to produce the division's strategic plan. The purpose of each strategic plan is to define how each unit will carry out relevant corporate goals. The planning may also be done by a *business area, strategic business unit,* and *strategy center*. These may be equivalent to a company's organizational element or its related lines of business grouped solely for strategic planning, and as such they may cut across formal organizational boundaries.

The president's or CEO's office integrates the plans of all corporate units by cutting here and adding there until it obtains a mosaic that best meets the needs of the company. For example, one area may be selected to grow, another to operate as a cash generator, and another to be phased out. It is within this mosaic of plans that a corporation's various technical groups plan and have their plans adjusted to overall corporate goals as presented by division, strategy center, or corporate managements.

The strategic plan is a road map for achieving goals.

Such integration of plans should involve considerable dialogue between a unit and the corporate officers before final selection of strategies that best fit the corporate goals and objectives. Typical questions that every unit will ask include:

- Where are we now?

  What is our status? Our competitors' status?
  What business and technical changes are occurring?
  What are the critical issues or concerns? (These are both internally and externally oriented.)

- Where do we want to be?

  What growth markets should be targeted?
  What are the market needs?
  What new products should we produce?

- How do we get there?

  How can we enhance our performance advantages?
  What are our best strategies?
  What resources do we need?

The strategic plan covers a 5- to 10-year period, although this period varies with the type of business. For example, a forestry related business looks farther ahead (10- to 15-year horizon) than electronics-related businesses do (a horizon of a few years at best).

The operating plan

Once each group of planners selects strategies that fulfill corporate goals, it drafts an operating plan. The operating plan (or profit plan) emphasizes the year (or fiscal year) directly ahead.

The operating plan is a blueprint for carrying out the strategic plan.

The operating plan includes:

- A definition of the business
- Technical and market trends
- Goals

- Technical constraints
- Needs and opportunities
- A review of the past year's results
- A financial accounting for the past year's results
- A review of the strategic plan
- Product termination
- A detailed analysis of R&D projects
- A financial forecast for the coming year

Thus, each entity's operating plan details current attempts by all segments of each operation to advance and return a profit.

Strategic Technical Planning

I turn now to a detailed discussion of strategic technical planning. The general outline of the method given here is based on the wealth of information on general strategic planning methods in the literature (a few of which are cited in the "References" and "Suggested Readings" at the end of this chapter). However, the content is not written for general management purposes but is directed specifically to the technical audience, and adaptation to the specific needs of your group is expected.

As contrasted to the two planning periods occurring in operations, R&D generally has one planning period. A planning team develops or redoes the R&D strategic and operational plan every five years (or other selected period). During other years it develops its operational plan, which includes a review of the strategic plan.

Many useful techniques are employed in the planning process, and Part 3 of this book provides information more detailed than that discussed here. If you have little knowledge of such techniques, it could be to your advantage to peruse the materials in Part 3 at this point. The references provide still other techniques and details.

It may seem obvious that R&D portfolios must be so aligned that they meet corporate goals, but too often R&D groups are left without the top management guidelines and collaboration required for a mutual definition of the project portfolios that are best for the company. Success depends on a dialogue between top management and the R&D leader on mission, goals, strategies, and means of implementation. If your superior doesn't open such discussions with you, you must initiate them. These are important aspects of participative R&D management. The relationship of corporate planning with R&D planning is shown in Fig. 3.1.

Top management sets R&D mission and goals.

As an R&D manager, you will fulfill corporate strategy by planning for change throughout the planning exercise. You will include uncertainties of innovation (probabilities of technical and market success) and uncertainties of the environment (effects of public policy, consumer mood, actions by the competition) in your considerations. You will recognize technology push (the brilliant idea seeking a market) and market pull (a market need seeking a product) and what the general corporate climate or attitude is on projects based on either.

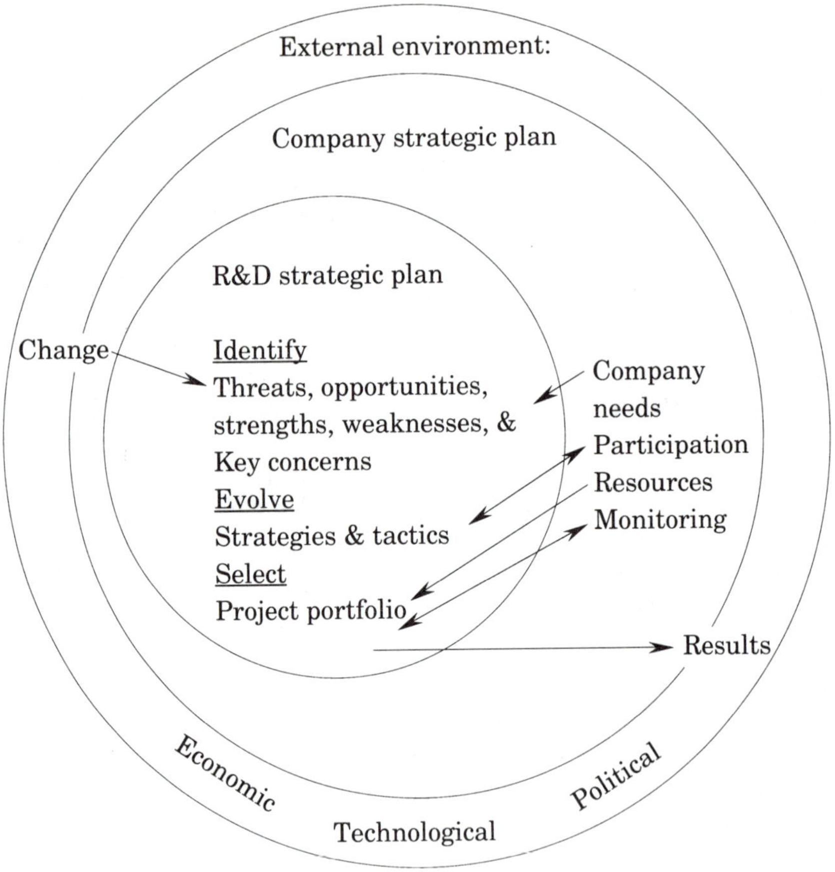

Figure 3.1 Relationship of corporate and R&D planning.

We live with uncertainty.

If you are manager of an operational R&D group, you will recognize the needs of the parent business unit; if you are in a central R&D group, you will recognize the needs of the corporation as a whole. In either case, you must have the means and the ability to monitor technology and to forecast change (see Chap. 7). A key requirement is to keep your eyes on horizons well beyond current technology. Also, you must recognize where and who the entrepreneurs and project champions are in your company.

Finally, top management must understand the point of uncertainty, be receptive to innovation, and be the stimulus for the process of strategic planning and the agitator for an innovative environment. If they do not, then your strategic technical planning will never evoke empathy and your group will flounder and fail.

An R&D group's managers form the planning team

The head of the R&D group in a business unit (a *unit* may be a section, department, division, combinations of these, or company) and the managers of the various R&D areas in the group should be the planning team members. How deeply the team draws its members from the organization depends on company size, commonality of interests among business units, and questions such as what is a reasonable team size. Ideally, senior professionals, managers of various functions, and planners in the business unit will assist.

The research managers form the planning team.

To simplify the following discussion, I have chosen a corporate R&D group as the planning group. In this example, the vice president of R&D is the team leader, with other members being managers of the group's various R&D areas. Managers of relevant operations will assist or be asked for assistance. Corporate officers and staff from selected functions may be asked to review critical points in the developing plan. (I find it necessary to harp on the need to form these associations early in the planning process. Sooner or later you're going to need them, and it's just good politics to have them in at the start.)

If you are the head of R&D and thereby the team leader, you *cannot* delegate the thought processes required for the planning process and the derivation of results. You can, of course, use every fact finding function available, but the team does the actual manipulation of inputs and produces the output results. Onerous? Yes, but once done, you will know more about your operation than ever before and be equipped to manage your assigned functions.

Good management does not delegate the planning process.

When the planning team is assembled, the team leader reminds team members of the unit's mission (also called *charter* or *definition of business*) and the R&D group's mission. If the team is in an operations unit, the mission statement emphasizes upgrades and means to advance market share of present products; if the team is in central R&D, emphasis is on new products, new technologies, and new opportunities. These mission statements are important, because they define the business and give its scope in clear, concise language. Two typical mission statements would be:

> Division X designs, manufactures, and sells reliable sensors and monitoring equipment to meet the severe environments within the mining industries. The mission of the R&D group is to enhance the performance of current products and to discover and develop new products that will aid in maintaining and advancing the market leadership of the division. In so doing, the R&D group will provide technical surveillance over current and emerging relevant technologies, monitor competitors' products and services, maintain and advance market share through upgrades and extensions to the product line, and develop selected new products within the scope of the business.
>
> Company X designs, manufactures, and sells hardware and services to the energy industries. The mission of the R&D group is to discover and develop products that will give the company commanding leadership in its selected business areas. In so doing, the corporate R&D group will provide surveillance over current and emerging relevant technologies, conceive and develop new products to meet future change, and provide problem-solving research and services to operations as needed.

Such mission statements focus the team on the issues important to the business of its parent unit.

The team leader presents the needs of top management to the team and discusses specifically the goals that management wants R&D to meet. (These needs should reflect, in part, the inputs from R&D, which has routinely routed its visions of the future to top management.) The goals of top management may be cast as general statements such as:

- Look at area X over the next few months and see if you can conceive an advanced method.

- Create a new generation of products in the near future from emerging technology A.

or as very specific goals such as:

- Provide division Y with an upgraded product Z using your materials technology, and let's see where you are in six months.
- Reduce materials costs of product B this year.

The planning team reviews any goals previously set out by the R&D group to determine their compatibility with current goals. The team identifies what can be done within available resources, what has to give, what employee and equipment resources are needed, and so forth. Goals are reviewed and refined or revised at the end of the strategy session. (A *goal* is usually defined as an achievement to be accomplished within a specified time period.)

Finally, the team discusses how to accomplish the six stages of planning (see the next section) and sets out tasks and schedules. The team leader also discusses the methods to be used in fulfilling the assigned tasks.

The strategic technical planning task required of the planning team can be facilitated by use of the six planning stages described in detail below. In the first stage, the team collects information. In the second stage, the team consolidates (categorizes, digests, and assimilates) this information into various lists. These lists are used in the succeeding stages of planning, and their comprehensiveness is critical to the planning effort. The next three stages are progressive refinements of the organized and developed findings. The stages of the planning process are given in Table 3.1.

A six-stage planning process organizes the planning efforts.

The six-stage planning process

Gathering information. The planning team's first task is to do a situation or position audit to determine the status of the company or unit. The team must know all about the company's products, and their performances, costs, ROIs, technologies, and market characteristics. In like manner, the team needs to know about competitors' product lines. Comparison of these facts may lead to identification of threats or opportunities. Sources of information are the team members' experiences, the experiences of colleagues in operations, internal and external reports, the literature, consultants, external conferences, and the company's marketing departments.

TABLE 3.1 The Six Stages of the Strategic Technical Planning Process

1. Information gathering stage
   - Determine status of the business unit
   - Ascertain needs of operations
   - Determine status of competition
   - Conduct technical planning studies
   - Consider key concerns and issues
2. Consolidation stage
   - Derive scenarios of possible futures
   - List needs, opportunities, threats, impacts
   - List key concerns and issues
   - List strengths and weaknesses
3. Strategy formulation stage
   - Analyze and evaluate lists of needs, etc., against list of key concerns and issues
   - Evaluate maturity of present technologies and possible use of new technologies
   - Match lists with strengths and weaknesses
   - Develop preliminary alternative strategies
   - Develop candidate tactics
   - Evaluate and suggest priority of strategies
4. Selection stage
   - Select one set of strategies, or
   - Look again at some new technology or combination of strategies and then select one set of strategies and tactics
5. Implementation stage
   - Consider project candidates (tactics) in depth
   - Test tactics against best and worst scenarios
   - Consider funding limitations
   - Suggest priorities of specific projects
   - Describe the group's R&D areas
   - Set goals
   - Draft strategic and operational plans
6. Review stage
   - Submit plans for review
   - Adjust plans as necessary

The team surveys the needs of the company or business unit as perceived by senior staff in the relevant operations of the unit. In this instance, the team ascertains needs in all the corporation's operating units. Also, it solicits input on new technologies that could help or affect operations. (In my experience, auditing the divisions' strategic and operational plans can be revealing, but I generally found that to get a real gut feeling I had to talk personally with colleagues in operations. You simply cannot expect an operating group to bare in writing to corporate officers a litany of dirty laundry. You have to meet with them on an informal, friendly basis to get a straight story. And then, of course, you must use discretion in how you handle such information, or you'll find a closed door the next time.)

Audit not only your company but also the consumer and the competition.

Although the team inquires about strengths and weaknesses from colleagues in operations, the team members should also consider from their own viewpoints what the company's strengths and weaknesses are, both in operations and in the R&D group. Similar information should be sought about competitors.

I stated in the preface the importance of new products and processes to corporate growth. The team must gather information on new technologies (technology push) and compare emerging technologies with the company's existing technologies. This will suggest how emerging technologies may affect the corporation, providing opportunities or threats. Technical forecasting may be necessary. These efforts, which I call *technical planning studies* (see Chap. 7), are usually done by the senior professional staff or a research manager. During such studies, technical people may also turn up opportunities by "imagineering" the adaptation of new technologies to current products or may derive new product ideas when considering the function or functions of current products. Questioning how such functions would be carried out without a knowledge of current products often leads to new ideas. To generate such ideas, tell persons not acquainted with your products, such as a new employee or members of a contract research house, about your needs and ask them to design a product without giving them access to conventional knowledge of current products.

Emerging technologies may yield opportunities or threats.

The team also should interview marketing departments and selected customers for needs of the marketplace (market pull), comparing such information with information of competitors' new product ventures.

During this period of information gathering, the team identifies key concerns and issues.[45] These may concern the relative proportions of short- and long-term research, the manner in which the research efforts of the different areas within R&D complement one another, and the risk attitude of management. The team leader may have to help by providing his or her best judgments on some of these points.

What are the key concerns about the research?

- Are we accounting for and giving correct priority to the needs of operations?

- How will the economic environment affect the company and the R&D content?
- Will our strategy match top management's needs?
- Is there leverage among the technologies (that is, reuse of technologies) or leverage by combining the talents of two research areas?
- What is competition doing?
- Do our (or their) products have some distinct advantage?
- Do we have appropriate employee and equipment resources to implement a specific strategy? The team should ask questions about these key concerns.

The efforts put forth in the information-gathering stage provide the basis for determining where the company wants to be.

Consolidating information. The team now consolidates the background material into lists of needs, opportunities, threats, strengths, and weaknesses. The types of lists needed are identified in Table 3.2. Although only one example is given for the content of each list in Table 3.2, the contents of these lists should be exhaustive. For example, the content of the "key concerns and issues" list is expanded somewhat in Table 3.3.

Digest the information and put it into usable form.

TABLE 3.2 Listings to Consolidate Background Material

Subject of list	Example of content
Needs	Component A is wearing at too fast a rate. Solution could lead to increased performance and an advantage over competition.
Opportunities	Laser cladding of A is not being used but could produce higher product performance at lower cost.
Threats	Company QRS is using laser cladding and threatening our market share.
Impacts	A cartel will form covering mineral B and may cause serious price increases.
Strengths	We have an exceptionally strong materials department.
Weaknesses	We have limited sources of mineral B.
Key concerns and issues	Can we develop a substitute and issues for mineral B? Should we?

TABLE 3.3 Expansion of Key Concerns and Issues List

Concern or issue	Comment
Does the product fulfill corporate goals?	What about growth rate and bottom-line impact?
Are goals set by management for R&D being met?	If not, why not?
Are goals set previously by R&D still valid?	Or have priorities changed?
Does the product have company relevance?	If not, how do we expect to make money?
Does it match the CEO's priorities and risk attitude?	What does the CEO think is important from R&D, and how much will he or she gamble?
Are there common interests among operating units?	Can the technology be reused by several units?
Is there an active interest from an operation?	We must transfer the project to a division that will champion and eventually sell the product.
Can the product be marketed?	Where is the marketing expertise?
Will the contemplated strategy support or buttress another strategy?	Is the technology reusable?
Are we using the newest or best technology?	Competition may be!
Will the customer perceive increased performance?	This is a chance to increase market share.
Will there be a significant effect on the bottom line?	The company must survive and grow.

Since the team makes decisions today that will affect the future, it needs a way to assess what the future may look like. Scenarios of possible alternative futures, derived according to the method[46] given in Chap. 7, provide narratives over a selected period of time under various themes. The team uses technical forecasting in the formulation of alternative futures and from these scenarios assesses possible impacts on the corporation. Needs of operations can be translated into threats and opportunities. The same is true of impacts from scenarios. So, the lists boil down to five: threats, opportunities, strengths, weaknesses, and key concerns.

The consolidation of information (its categorization under the headings stated in Table 3.2), if exhaustively and thoughtfully done, permits the planning team to derive alternative strategies in the next planning stage. This categorization also helps identify "holes" in their information, new information that must be obtained, and additional forecasting needed to assess opportunities or threats. For any new product or technology opportunities identified, the team assesses the possible benefits these may give.

Formulating strategies. The preparation of lists in the previous stage is a simple way of categorizing considerable data and thereby giving consideration to all data. The planning team can manipulate such lists by matching opportunities and threats with strengths and weaknesses to derive a list of preliminary alternative strategies. (A *strategy* is an action, formulated from analysis of environmental conditions, giving direction toward fulfilling a goal.) These strategies are subsequently evaluated to select those best suited to answer goals within the allocated resources. In this stage, workshop and brainstorming sessions are especially useful to team members. "Thinking aloud" permits team members to see how the ideas of some members complement those of others. Group thinking is a very useful tool in firming up or shooting down strategies.

Obtain strategies from the derived lists.

As an example of using these lists, suppose an internal need is improved materials, which if met would represent an opportunity to expand market share. However, this may not be met easily, because of an internal weakness such as lack of materials expertise (which lack has been listed). One possible approach could be to use consultants and (or) contract research and, if some positive results are obtained, to hire the necessary expertise to follow through. With this approach, a generic expertise would be built which would be important not only to the identified need but also to the possibility of expanded applications in other areas of the company.

An internal need met by an internal strength (again, evident from the listed information) could yield more specific answers, for example, meeting a wear problem with coatings of microcrystalline alloys.

An external threat or opportunity identified on the list might be matched with a strength (or weakness) of one of the listed capabilities, and the combination might suggest specific action to deal with the threat or take advantage of the opportunity; that is, a strategy has been derived. If, for example, competition is laser welding and producing a better product, the company had best go to laser welding, decide to default the business, or rethink the product and either eliminate the need for the process or leapfrog this technology with a still better one. Whatever the decision, another action or strategy has been identified and added to the list of alternatives.

Opportunities or threats plus strengths or weaknesses suggest a list of preliminary strategies.

The planning team must relate, without bias, each listed opportunity or threat to the company's ability to obtain results and add these conclusions to the list of alternative strategies.

They should compare the company's strengths and weaknesses with the strengths and weaknesses of the competition. Additional threats or opportunities may then become evident.

So far, the team has made few judgments on the relative values or merits of these preliminary strategies. As they derive preliminary alternative strategies, the team should also consider candidate tactics or projects that might lead to the results desired. From special company capabilities certain candidate projects become obvious; however, the team may need some new idea to obtain a significant product advantage. Discussions with selected staff members may help develop such an idea, or the team may have to let the subject "cook" until the "right side of the brain" provides the creative idea on which to base a project. (Candidate projects will be reviewed in depth in a later stage.) The task now is for the team to compare and evaluate alternative strategies that will eventually result in a project portfolio.

Start to consider tactics and needed resources.

It is important that the team obtain top management's guidelines[47] and the benefit of their thinking before beginning the creation and evaluation of alternative strategies. Any delay on management's part in providing the needed guidelines would be wasteful and incompatible with the establishment of an appropriate climate for greatest productivity. It is the team leader's duty often to remind the team of corporate goals and their original discussion of goals.

The strategy formulation stage establishes a preliminary list of alternative strategies with some candidate tactics by which the company can get to where it wants to be.

Know where the company wants to be so that strategy selection can be meaningful.

Selecting strategies. All the work accomplished in the first three technical planning stages leads to a selection of strategies that best support corporate goals within allocated resources, the task in this stage. The team members evaluate preliminary alternative strategies against the information they have listed under key concerns or issues (illustrated in Table 3.3). They examine strategies to see if any can be questioned based on information in this list, balancing those strategies reflecting near-term company needs against those reflecting long-term future opportunities. They discuss and list candidate means of carrying out the strategies, and begin to pare down the number of alternative strategies when these comparisons reveal incompatibilities. They test strategies and candidate tactics against friendly "outsiders" in the company for possible improvements and for barriers within the company organization.

The last two items in Table 3.3 are very important and can be paraphrased as "performance leads to market share which leads to profitability." The Strategic Planning Institute has shown that ROI increases with market share.[16] The institute has also shown a strong connection between new-product sales and growth in relative market share.[15] Building market share with new products, of course, costs money in the short term. To reduce such costs, the institute suggests that companies which are winners are those which introduce new products from a position of strength (that is, successful companies have either a broader product line than the competition does, a higher level of product quality, a stronger competitive position, or all three). This means that combining strengths with opportunities or threats produces a strategy with a greater probability of success.

Innovations help to increase market share and profitability.

However, a warning is needed: You must be very careful in applying such experience across the board and thereby suppressing innovation. Many major innovations have been generated *outside the industry being changed.*[48] A good example of this was the introduction of transistors, which were marketed first by Texas Instruments, not by RCA or GE, each of which had a base in vacuum tube technology. Thus, a technology push may be riskier or take a longer time (or both) than an alternative market pull project, but it may well provide the more handsome reward. Implementation of selected strategies may mean significant change in some operating units, so it is important at this stage to obtain these units' active interest. This might be done by forwarding a statement of the strategy and a description of a possible project to the division that would be affected and follow up on the telephone to discover its interest. Cooperation is readily obtainable when there are needs to be met or threats to be faced; it is with new opportunities that problems arise. Change is agonizing, producing problems in both marketing and manufacturing. The key is for R&D to generate change in such a manner that the concerned unit can and will accept the change. When a new product is being considered, it is helpful to locate a possible champion in the relevant division, talk to the marketing group, and have discussions with division top management about the product and the changes it could bring. Another gauge of the potential success of the team's selection of strategies is the probability of success in a later technology transfer (more on this later), which will depend on whether discussions with operations indicated they were enthusiastically for, neutral, or against the given strategy or tactic.

Change becomes acceptable when new profits are seen as attainable.

I want to emphasize that the team leader and (or) the relevant manager should discuss any proposed project with the concerned operating group, and the group's top manager should be present. Such discussion should include not only the consequences of the project's failing but also of its succeeding. This may sound peculiar, but any operating group, while welcoming "free" research, may have fits of agony when the research successfully ends in a new product and they must buy an expensive piece of capital equipment to cope with success. Oh, that bottom line!

Discuss the consequences of project success.

Although it is important to attain operations' interest, you must get it in such a way that you obtain an objective commitment. Many times you can expect the arrogance of success or the syndromes of "not invented here," "that will never work," or "we can't afford to change." Give careful thought to trying to establish a relationship whereby your ideas receive a fair and objective hearing.

Throughout the planning exercise, each manager should continually review progress with his or her staff. Even though each manager may not get 100 percent concurrence with his or her contributions to the plan, the result will be 100 percent support for those that are accepted, because of the participatory leadership that has been established. Of course, it can get sticky when a decision is made by the planning team, the head of R&D, or corporate management to constrict, reduce, or limit one group so that another can enlarge its efforts. In such a case, the vice president of R&D must convince the staff of the virtues of the proposed action.

The selection of strategies is an iterative process; the team should go over these strategies and candidate tactics many times, improving some and culling others to arrive at a list of compatible strategies that may fulfill the goals set forth. This set of strategies must also correspond to the probable resources to be made available. This can be ensured by considering the corresponding tactics (projects) and the probable costs of R&D—the task of the next stage.

Iterate and cull, then improve the balance.

Implementing strategies. Having selected a set of strategies and some candidate tactics that may best fulfill them, and having obtained an indication of probable interest from operations, the team undertakes an intensive, in-depth exploration of possible tactics to derive the final tactics for accomplishing its goals. Considerable time and effort must be expended to identify all possible tactics, select the most reasonable, and, where ideation is lacking, allow the staff to propose a tactic at a later date.

The team then puts strategies in order of priority, primarily by comparing the probable costs of research and development and the potential benefits to be gained if a product is commercialized. This is not an easy task, nor will the numbers be accurate at this point; the team's past experience will help. To get costs, it considers the possible technical, legal, and market constraints and the program to overcome them. Then it considers benefits. If certain materials and processing techniques can yield new products, it estimates the best combinations of materials and design parameters that will supply these products and identifies those products that will have the greater bottom-line impact on the company. Questions to be answered include the following: "What product advantages might accrue, and will these advantages lead to a reasonable market penetration?" "Does the product fulfill user needs?" "What market share can be expected?" Some techniques given in Chaps. 7 and 8 will be useful.

Establish priorities.

The team uses decision analysis to determine benefit-to-cost ratios, evolving additional or alternative projects which are subject to the questions and evaluations stated above.[49] The team repeats these considerations for each strategy on the list, and establishes priorities by comparing benefit-to-cost ratios, evaluating operations' reaction to the concept, ascertaining the team's judgments on how the strategies match up with the key concerns and issues, and estimating the probability of getting a successful product to market on time and within the cost target.

Such measures are subject to a numerical evaluation or a score chart methodology, but keep in mind that if the projects are not enthusiastically embraced by both technical and management personnel in the operating units, such mathematical ratings will probably be worthless. Don't let such measures be a substitute for thinking.

Evaluate numerically, then use common sense.

I want to acknowledge that exceptional case where a project is initiated by corporate R&D because of its remarkable potential in spite of no interest in active participation by any of the operating divisions. Such a project will be successful only if top management recognizes its possible value and supports the commitment of resources, if R&D management monitors it closely, and if communications between corporate and R&D managements include frank and clear appraisals of progress.

The R&D planning team must now match total project costs with the probable budget. Assume that the team has greater project costs than it knows will be acceptable. Team members have to retain those strategies that will have the greatest impact on, and best fulfill, corporate goals, and they have to defer the balance. If the team wishes to expedite one area of research without expanding the R&D budget, then another research area will have to participate by deferring some of its own R&D. If there is an exceptional project requiring an unusually large funding, the team should separate it from the usual portfolio and request special funding based on its peculiar opportunity. Depending on your company policy, special funding could come from the company, government contracts, special venture groups, or a venture with a second company.

Yes, the portfolio must be within the probable budget.

Although priorities have been set by the above techniques, sometimes additional means may be needed for delineating a final project portfolio. In such cases, Owen[50] presents an interesting mathematical method. With this method, the potential value of each project is multiplied by its probability of technical success, and R&D costs subtracted to get expected values. The projects and the total project costs are then juggled to select those projects which will maximize the total expected values to be obtained within the anticipated budget. These projects may then be examined on a 2 × 2 matrix, with the probability of technical success plotted against net present value. The locations of the projects in the four quadrants give a visual depiction of their values to the company and permit selection between near- and long-term opportunities. (Please refer to the decision analysis section in Chap. 7.) The end result is a project portfolio dedicated to answering corporate goals within allocated resources.

Compare expected values with project costs to get the most for your dollar.

Technology Futures* offers another approach to identifying, analyzing, and evaluating alternatives. This is a software package called Techmate and is described as integrating forecasts on costs, development times, technical parameters, rates of adoption, probability of success, and external events into the decision-making process. The output will aid in selecting the most beneficial strategies and tactics.

The last planning task for the team is to review and restate goals in terms that permit monitoring subsequent performance toward meeting them. There are two parts to a goal statement: what is to be accomplished and the time in which it is to be completed. At this point, the team should also ask "Do these specific goals include those set forth by top management?" (If not, the team had best have another iteration.) "Do they include goals previously set?" (If not, the team should reconcile the differences.)

The selected strategies lead to selected tactics, and these form the project portfolio. When the projects are grouped, they form the main research and development areas (or thrusts) of the R&D group (for example, materials, energy production, energy conversion). These R&D areas are summarized in a brief statement.

As a result of accomplishing the tasks identified in the first five stages of the technical planning process, the team has:

- A set of strategies
- Corresponding tactics in order of priority
- A set of specific goals
- A global statement on the R&D areas of the group

At this point, the vice president of R&D meets with management for an informal discussion about the results of the planning exercise (informal discussions from the start of the exercise would be advantageous, but one at this point is mandatory to prepare management for the results of the planning exercise and to obtain knowledge of any reservation which can be answered in the upcoming planning report). Although the meeting is informal, have visual aids such as placards

*Technology Futures, Inc., 6034 Courtyard Dr., Suite 380, Austin, TX 78730.

showing summaries. Allow for dialogue. Seek opinions. Don't let a reservation go unanswered. Probe to find basic reasons for any reservations, and then seek resolution. When the results of the meeting are evaluated, the team is ready to write the final draft of its strategic and operating plan.

Each research manager on the team, who has developed an operating plan for the research for his or her area, puts the plan into final form. Projects are listed in order of priority, and information is presented, such as a Gantt chart of projects versus time, project costs, a personnel allocation schedule, and review dates. All area inputs are combined to show support for the strategies developed to fulfill company goals, and the result is the R&D group's final strategic and operating plan.

This plan should be presented to top management in fewer than 50 pages, including introduction, recommended strategies and tactics, benefits, implementation means, goals, cost estimates, research area summaries, and a background synopsis. The presentation can be supplemented by an unexpurgated report with all background material, research area reports, and so on. Some say that a strategic plan should be capable of presentation in fewer than a dozen pages. The shorter the better, so long as it communicates the results.

An effective report will be short and succinct.

Reviewing the plan. The head of R&D presents the plan and its backup document to the president or CEO, presumably with an oral reporting of its sagacity, benefits, and great opportunities. If the president or CEO has participated in some of the discussions on alternative strategies and candidate tactics, this is a tremendous help to the review. However, top management must then view the plan as it relates to corporate needs, risks, and available funding. If management's goals have been met and if other goals are consistent with the unit's mission and are within the anticipated funding, the plan should find ready acceptance.

As the environment changes, as new technology appears, or as competition adds some surprises, an ad hoc review of both the strategic and the operating plans is warranted. Also, new ideas can be referenced to the strategic plan, making any decisions on actions easier.

Environmental changes may force plan changes.

As said earlier, the planning team redoes the strategic plan every five years (or some other selected period). The team also redoes the operating plan in intervening years.

The annual operating plan includes:

- An evaluation of technical and market trends by each manager
- A discussion of the importance of any emerging technology as determined by selective technical planning studies
- A review of the needs of the operating divisions
- An identification of any new threats and opportunities
- Any proposed changes to the strategic plan
- The project portfolio as reviewed and modified by each manager
- Supporting cost and benefit data for each project
- A budget for the ensuing year

Sources of information that aid in identifying technical and market trends are given in Exhibit 3.1.

Exhibit 3.1 Sources of information on technical and market trends

Cambridge Energy Research Associates, 56 John Kennedy St., Suite 5, Cambridge, MA 02138.
Research and consulting specializing in analysis of current energy issues and basic energy trends.
Compressed Air Magazine.
"This and That" is a regular column on new technology.
Control Engineering.
"Control Business Directions" is a regular column on technology and marketing.
Crane Today.
"Overseas News" is a monthly section.
Department of Energy (DOE), 1000 Independence Ave., S.W., Washington, DC 20585.
Provides technical information.
Dialog Information Retrieval Service, 3460 Hillview Ave., Palo Alto, CA 94304.
Provides computer search service for technical and market information.
Dun's Marketing Services, Three Sylvan Way, Parsippany, NJ, 07054- 3896.
Access to over 8-million businesses and professionals in the U.S. and over 800,000 companies internationally.
Frost and Sullivan, Inc., 106 Fulton Street, New York, NY 10038.
Provides market studies.
Iron Age, January issues.
Each January issue has an "Annual Forecast" section.
Laser Focus, January issues.
Each January issue has a "Review and Outlook" section.
McGraw-Hill Publications Company, Economics Department, 1221 Avenue of the Americas, New York, NY 10020.
Publishes business and economic trends.
Metal Progress, January issues.
Each January issue has a "Trends in Materials" section. National
Bureau of Economic Research, 261 Madison Ave., New York, 10016.
Economic facts on selected subjects.

National Science Foundation, Washington, DC 20550.
Publishes "Annual Science & Technology Report" and "Five-Year Outlook on Science & Technology." Also, "Science Indicators," which provides quantitative data about recent trends, such as investments in and resources for science and technology.
National Technical Information Service (NTIS), U.S. Department of Commerce, 5285 Port Royal Rd., Springfield, VA 22161.
Publishes technical information.
Pergamon Press, Inc., Fairview Park, Elmsford, NY 10523.
Its "Highlights of the Literature" series gives an overview of outstanding research, abstracts from the literature, and an extensive bibliography.
Predicast, 1101 Cedar Ave., Cleveland, OH 44106.
Publishes technical and market information.
Prestwick Publications, P.O. Box 205, Burnt Hills, NY 12027.
Publishes technical and market information.
Society of Automotive Engineers (SAE), 400 Commonwealth Dr., Warrendale, PA 15096.
Publishes technical information.
Technical Insights, Inc., 158 Linwood Plaza, Fort Lee, NJ 07024.
Publishes annual reports on research and development.
Technology Forecasts, published monthly by PWG Publications, 205 S. Beverly Dr., Beverly Hills, CA 90212.
Provides information on technical trends.
U.S. Bureau of Economic Analysis, U.S. Department of Commerce, Fourteenth St., N.W., Washington, DC 20230.
Publishes annual "Industrial Outlook on 250 Industries with Projections Four Years Ahead."
U.S. Department of Commerce, 5285 Port Royal Rd., Springfield, VA 22161.
Advisory group for aerospace R&D. Publisher of "A User's Guide" (AD-780 061); AGARD Lecture Series 69, "How to Obtain Information in Different Fields of Science & Technology."
Value Line Inc., 711 Third Ave., New York, NY 10017.
Publishes technical and market information.
Welding Design and Fabrication.
Has a monthly section, "Industry News," on marketing and technology.

(Continued)

The planning process accounts for all appropriate interrelationships

The R&D group's managers form the planning team. The vice president of R&D obtains the mission statement and statements of goals from the group's superior and communicates them to the planning team. (Sometimes the vice president of R&D must fabricate the mission statement and formulate goals based on previous informal discussions and then obtain approval from the superior!) The vice president also discusses with the planning team any prejudices, biases, noteworthy attitudes, and special requests of management, including corporate operations vice presidents and division presidents.

The R&D group's managers obtain input from and provide feedback to their staffs. They also discuss strategies and tactics with downstream groups to obtain interest and inputs. The downstream groups include engineering, manufacturing, and marketing. Support groups probably have the least input, but their participation can be valuable, especially with budgets and personnel.

To get a home run, you have to touch all the bases.

Goals, strategies, and tactics are the output of the team, and the decisions represent the participation and collaboration of all role players in the innovation process.

> Thus, the planning process has the planning team touching all bases, and innovation's management circle has been recognized and used.

Summary

- Strategic planning should be conducted corporatewide, with each unit's plan integrated into a mosaic that becomes the corporate plan.
- Strategic planning develops alternatives based on forecasts of the future.
- The strategic technical plan must be integrated with the corporate plan.
- The technical planning team is made up of the leader of R&D and his or her managers; the team's delegation of responsibility is only possible for data accumulation.
- The planning process is a six-stage process that includes information gathering, consolidation of information, strategy formulation, strategy selection, implementation through tactics formulation and selection, and review by top management.
- The process is as follows:

```
Top management → Mission, goals → Strategies
   ↑                                    ↓
Plans ← Thrusts, goals, portfolio ← Tactics
```

- A major difference between strategic technical planning as presented here and strategic planning for general management purposes is that tactical planning is not a separate activity but, rather, starts during the strategy formulation stage. Strategic technical planning yields a strategic and operational plan from the one planning period rather than separate strategic and operational plans from the two planning periods typical of management planning. In years that strategic technical planning is not the major task, the strategic technical plan is reviewed during formulation of the annual operational plan.

Cases and Incidents

The Position Audit

Dr. Watts has for some time been aware of the lab's lack of management experience. Dr. Helms, head of the lab, has previous supervisory experience, but neither he nor any of his managers has had much industrial management experience. Roy, who is manager of fundamental alloy investigations, is the key professional behind existing efforts but is more interested in personal achievement than power and is basically a retiring person. Larry, who is manager of applications development, is an excellent engineer and so agreeable that he has trouble saying "No" to anyone. Tim is manager of analytical investigations and an analytical specialist. No one in the group is a ball of fire.

Dr. Watts asks Dr. Helms to call the managers together for another meeting. He wants, at the very least, to find out where they stand. He asks them about the costs and benefits of their current efforts and which individuals in operations show interest. The answers suggest many benefits (none of which have been quantified) and the interest of everyone (none of whom has invested in the effort). Not many specifics but many generalities. Much enthusiasm.

Dr. Watts says that the group needs specifics about their program and what it can do for the company. He tells Helms and the staff that he would like to get the following information:

- Properties of their new materials and those of the competition.
- Applications of the new materials to current products, giving benefits.
- Possible new products that fit the businesses of the company.
- Practicality of current processes.
- What the divisions know about the work and whether they need more information; whether they anticipate viable products; whether they have capabilities to follow through on the work; whether they will follow through on development, if prototype testing is favorable.
- A summary of threats and opportunities possibly answered by these materials.
- What communications exist with downstream groups and how can closer working relations can be obtained.

- What the R&D group knows about decision and profitability analysis.
- What market analyses have been done and what should be done.
- Patent opportunities and conflicts that exist.
- Cost-benefit ratios, rates of return, and internal rates of return for their products and also for current products on the market.

The staff gets busy over the next few days. They are immediately aware of what they don't know, which is something of a start. There's some wasted motion, but they soon start to settle down and get answers to Watts's questions.

Tim asks if all this is really necessary; it's a lot of work. Watts says it's necessary if you want reasonable conclusions about whether your area can be profitable, and whether changes should be made in personnel or procedures (for example, whether Tim's job is really necessary!).

Watts ends by saying, "You get me the information I ask for and we'll know what our position is. From that, we'll know what we have to do to get this show on the road."

Questions

- Are there other investigations that should be made?
- Shouldn't Dr. Watts have personal discussions with the staff on a one-to-one basis? Discussions with the CEO? With selected division personnel?
- What is the team going to do with the information it gets? Is there an optimum way to handle it?
- What is Helms doing to manage the work in an expeditious manner?
- What is the staff learning about the company and its competition? (See Watts's list.)

The Strategic Plan

After discussion with the CEO, Dr. Watts and Dr. Helms conclude that the CEO has no ideas about what the R&D group should be doing beyond his original objective. The CEO feels that the company ought to be doing more to be a leader in its fields. There is no doubt in his mind that new businesses exist which they should be in and that a corporate laboratory could have that mission. Watts and Helms obtain agreement from the CEO that the R&D group will draft a long-range plan for his review and consideration. Dr. Helms, the managers (Roy, Larry, and Tim), and Watts form the planning team. Watts' role is to instruct in how to plan.

Their first step is to read policy and philosophy statements issued by the corporation, read the annual reports of all the divisions, and gain a clear idea of what the company businesses are.

Watts says they must remember that a corporate R&D group's

role is not only to emphasize long-range, new product research but also to get involved in upgrades and some service work. Also, deriving a mission statement and goals is an important first step (and is seriously hurt if there is no input from topside).

From what Watts has learned from the CEO and what the team has learned from readings, several goals are identified. Helms and Watts then interview selected division personnel to determine their current and future plans, their problems, any interesting new technologies, competitive products, and their strengths and weaknesses.

Roy and Larry do special technical planning studies on areas of interest to the company in which emerging technologies might permit some sizable advances. Selected professionals in each of the three R&D areas are asked to identify market and technical trends of the company's businesses from the literature, outside research services, and the divisions.

Everyone on the team compiles lists of strengths, weaknesses, opportunities, and threats and notes any concerns, questions, or issues they have as their data gathering proceeds. Roy's particular concern is about the amount of short- versus long-term R&D the CEO would authorize, and Larry's concern is the acceptance of R&D programs by downstream groups. Tim wants to get involved in the mathematics of project and portfolio analysis. Dr. Helms seems a bit overwhelmed and leans heavily on Watts' suggestions on how to handle the data.

Watts tells the group they must match strengths and weaknesses with opportunities and threats from data in their lists and develop a list of alternative strategies. He says that to derive a strategy, one must know what the word *means* in *all* its ramifications. Then, taking into account the key concerns and issues, they must start to pare down the list of strategies.

Everybody is getting interested, and Watts says, "Now we're getting somewhere. Let's make a selection of strategies and consider the possible tactics to fulfill them. Alternative strategies give us room to make a final selection that will fulfill company goals, and, of course, strong tactics are needed or we don't go anywhere. With a plan, you know where you're going and have the strategies to get there."

The team also has to rationalize how the group can grow over a period of years (say, five). Growth will depend largely on the team's ability to get good professionals and on everyone's feel for probable budgets. Within these probable time and cost parameters, the team selects strategies and tactics and redefines goals. Tim (with Watts's help) does some decision analysis as an aid in selecting the projects that will optimize benefits for the estimated costs. Helms sends the resulting plan to the CEO.

Helms later says, "Good God, it's accepted!"

Watts replies, "What did you expect? We did our homework."

Questions

- Would you do things other than or besides those mentioned?
- How specifically can you state your company's business?
- Do you know what a mission statement involves?
- What is a strategy? A tactic?
- How would you establish priorities among strategies and tactics?
- What degree and sorts of exchange should managers have with staff?
- Why hadn't this exercise been done long before this?

The Team: Shock Tactics

It is finally obvious that the corporate R&D group needs more firepower. A position is found for Dr. Helms in one of the divisions, and Dr. Shock, a manager of a group similar to that planned, is brought in. Dr. Shock has had management experience over a 100 people at his previous position, has many management courses under his belt, and has had many dealings with production people.

Dr. Shock makes Roy a senior scientist and replaces him with Dr. Brown, who has extensive metallurgical experience and is known to Shock through past professional meetings. An engineer from one of the operating groups comes in to work with Larry on the group's initial product.

Other personnel additions will be requested as noted in the strategic and operating plan if the group obtains results according to plan.

Dr. Shock makes an immediate difference. He calls the staff together, goes over the plan, and talks about how he expects it to be implemented. He initiates weekly meetings with the managers and tells them to arrive each Monday morning at 8 a.m. prepared to discuss results obtained versus the previous week's plans, problems, market events, new technologies, and next week's plans. He compliments Roy on his technical advances and says that he expects exciting new results under his new charter. He tells Larry to give every consideration to suggestions from the new person (the engineer from operations) and states that any potential conflict is to be brought to his attention immediately.

Dr. Shock also initiates brief monthly reports to the CEO and special reports to the participating divisions. Shock instructs his managers to start market research as appropriate and as approved by him.

Watts says, happily, "This place is beginning to buzz. I sense a fresh climate, one that is conducive to achievement."

Questions

- Should Shock be doing anything else?

- What about creativity?
- Idea review?
- What about adequacy of facilities and equipment?
- What should the weekly reports contain? The monthly reports? The quarterly reports? The annual reports?

Staff Development

Shock reviews what has been done to date on staff development. To implement the strategic plan, he has enlarged the staff to include an organic coatings group to parallel the alloy development of Dr. Brown. Dr. Chen heads the new group.

Because much of the company's hardware has electronic controls and the trend is toward hardware with a substantial electronics portion, Shock starts a small electronics group and brings in Mr. Click to initiate the effort. Harry is transferred from Larry's area to support Click's efforts.

Also, the company, which uses tonnage quantities of chemicals for corrosion control and lubrication, decides to produce its own line of chemicals and to develop a line of water-treatment chemicals. A chemist, Dr. Cory, transfers from a division to head this effort.

Shock tells Watts that during one of his early visits to the corporate offices he met most of the support people in accounting, finance, and personnel. In personnel, he found a staff development group that has various programs for employees at different levels of management. As a result, Leonard, a senior chemist in Dr. Chen's group, attends an advanced technical course at the local university and will have his tuition reimbursed by the company. Larry attends a management course.

Watts mentions that the staff must be made aware that applications for reimbursement of the costs of courses will be considered and approved according to the extent that they benefit both the person and the company and that there is a limit to what can be done each year.

Shock also tells Watts that he appointed Dr. Brown as the first seminar chair, with rotations to occur annually. The seminars include the existing staff (to give experience in oral presentations and to cross-fertilize the group on what everyone is doing), university professors (to provide input on new science and technology), and representatives of operations (to extend the group's knowledge of company business).

Shock asks, "What else should I do?"

Watts tells him he has made a good start on implementing the strategic plan. He also warns Shock that with new people coming into an established group there is always the question of equitable salary and benefits, especially for the existing staff. He suggests that Shock go over staff salaries carefully, discuss his findings with personnel, and iron out any inequity. Watts knows that Shock has kept salary ratings in mind when hiring, so he doesn't anticipate any big problems, but it's best to be sure.

Questions

- Is the salary schedule for your company correct for a mixture of professionals and technicians?
- What about establishing a periodic meeting with a representative from personnel to review salaries and any personnel problems?
- How about establishing performance awards?
- Is it clear to your staff that setting and meeting goals is an appraisal item?

References

42. F. W. Gluck, S. P. Kaufman, and A. S. Walleck, "Strategic Management for Competitive Advantage," *Harvard Bus. Rev.,* July–August 1980, pp. 154–161.
43. S. H. Hoch, "Integrating Social, Economic, Political, and Technical Forecasts into Business Strategy," Industrial Research Institute, New York, May 1981.
44. P. B. Skov and C. S. Spetzler, "Strategic Management Process," *Business Intelligence Program,* Res. Rep. 617, SRI International, Menlo Park, Calif., August 1979.
45. W. S. Royce, "Generating Strategic Alternatives," *Business Intelligence Program,* Res. Rep. 620, SRI International, Menlo Park, Calif., November 1979.
46. J. H. Vanston, Jr., et al., "Alternate Scenario Planning," *Tech. Forecasting and Social Change,* **10**:159–180 (1977).
47. Technical Insights, Inc., "R&D's Role in Writing a Product Innovation Charter," *Inside R&D,* Sept. 8, 1982, pp. 3–4.
48. R. Charpie and D. V. DeSimone, "Technological Innovation: Its Environment & Management," U.S. Dept. of Commerce, Washington, D.C., January 1967 (reprinted September 1967).
49. J. M. Harris, R. W. Shaw, Jr., and W. P. Sommers, "The Strategic Management of Technology," *Outlook,* Fall–Winter 1981, pp. 20–26.
50. D. L. Owen, "Selecting and Managing Your Research Portfolio," paper presented to the International Congress on Technology Exchange, Pittsburgh, Pa., May 3–6, 1982.

Suggested Readings

Bacon, F. R., Jr., and T. W. Butler, Jr.: *Planned Innovation,* 2d ed., University of Michigan, 1981.

Badawy, M. K.: "Applying Management by Objectives to R&D Labs," *Res. Mgmt.,* October 1976, pp. 129–133.

Birnbaum, W. S.: "Strategic Planning: How to Get Started," *Bus. to Bus.,* April 1985, pp. 38–39.

Booz-Allen & Hamilton: "New Products Management for the 1980s," New York, 1982.

Day, G. S.: "Diagnosing the Product Portfolio," *J. Marketing* **41**(2): 29–38 (1977).

Ellis, D. J., and P. P. Pekar, Jr.: *Planning for Non-planners,* AMACOM, New York, 1980.

Industrial Research Institute, "Planning Tools for Effective Research Management in the 1980s," Dec. 15, 1981. See articles by D. B. Merrifield, "Project Selection in the 1980s"; S. H. Hoch, "Integrating Social, Economic, Political, and Technical Forecasts

into Business Strategy"; D. Bitondo, "Linking Technological and Business Planning"; and H. Tecklenberg, "Using Social Scientists in R&D."
O'Conner, R.: "Preparing Managers for Planning," Report 781, The Conference Board, New York, 1980.
Ruggles, R. L., Jr.: "How to Integrate R&D and Corporate Goals," *Mgmt. Rev.*, September 1982, pp. 8–17.
Steele, L. W.: *Managing Technology: The Strategic View,* McGraw-Hill, New York, 1989.
Steiner, G. A.: *Top Management Planning,* Macmillan, New York, 1969.
———: *The "How" of Strategic Planning,* AMACOM, New York, 1978.
Vancil, R. F.: "Strategy Formulation in Complex Organizations," SMR Portfolio for Innovative Managers, Sloan Management Review Association, MIT, Cambridge, Mass., 1981.
Weil, E. D., and R. R. Cangemi: "Linking Long-Range Research to Strategic Planning," *Res. Mgmt.*, May–June 1983, pp. 32–39.

Chapter

4

Implementation Activities

Questions answered in Chapter 4

- What are some of the ways in which an R&D group can be organized?
- What is a task force?
- Why are a creative staff and environment important? What kills creativity? What promotes it?
- How can managers best use the talents of a group to solve problems?
- What is the test of feasibility for an idea?
- What are the characteristics of a potentially successful idea?
- What is the need for and how do we deploy staff, facilitators, and consultants?
- Why must constraints be recognized early?
- Why must managers think early of technology transfers?

As an R&D manager your primary concern is with the first three phases of the innovation process—ideation, feasibility, and research and development. A significant part of your group's work is matching needs and opportunities with ideas for new products and processes. Your involvement with engineering, manufacturing, marketing, and sales is in the area of technology transfer and in keeping these functions aware of, and involved with, what you and your research group are doing to carry out the company's R&D program. And they, of course, may come back to you with problems about specific products or processes.

There is much more to implementing the program plans than creating and developing ideas to provide a new product concept. If it is a new group, organization may be a first step. Staff creativity and a creative environment for the staff are among a manager's constant concerns. The orderly development of a product or process concept eliminates poor ideas early and allows for the development of good ideas in a cost-efficient manner. "Orderly development" includes the selection of projects that fulfill corporate goals and the selection of staff and consultants who are "best fits" for each project. Also, the R&D group's alertness to problems possibly arising in eventual technology transfers and its need to account for the requirements of downstream groups during current investigations are forward-looking activities that help to ensure success through all stages of the innovation process.

There's more to this business than meets the eye.

This chapter starts with a discussion of corporate organizational structure and goes on to the organization of the R&D group and such topics as fostering of creativity, initiation of projects, deployment of people, recognition of constraints, and actions necessary in the transfer of technology from R&D to engineering and manufacturing. I approach this discussion from a practical perspective, giving examples from past experiences to illustrate various points.

Organizational Structure

Although I will not dwell on the organization of companies, an understanding of the most common and elementary variations in company organization leads to a better understanding of the relationship of R&D groups to the balance of the company. Company organizations can use either a pyramidal or a flat hierarchy, as shown in Figs. 4.1 and 4.2, respectively. The pyramidal hierarchy, usually a centralized organization, has many levels of management. A manager's span of control is small, and the number of layers to transcend vertically often damps communications. The flat hierarchy adapts to decentralized organizations and has a small number of managerial levels. A manager's span of control is large; vertical communication is easier than horizontal communication.

Companies may also organize by product line or by function. Product A, B, C, and D groups report to the president, and each product line has its own marketing, sales, finance, international, and R&D functions. (This would make it a pyramidal structure.) Or various product lines may be combined to form an operations group reporting to the president, with each group having its own group of functions. Geographic separation usually means a separate product line per region and probably necessitates a flat, decentralized organization. In such instances, each geographic location has operational R&D as one of its

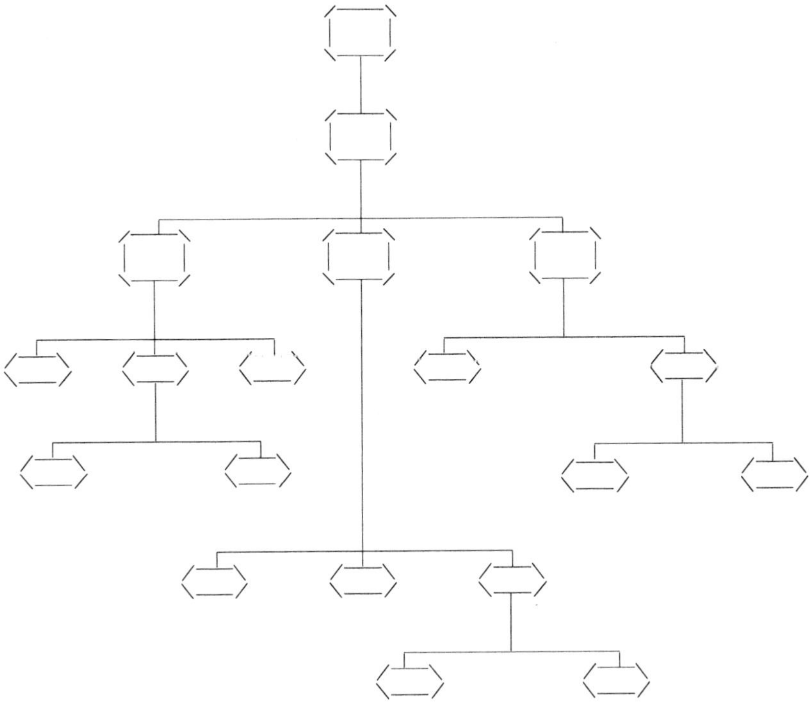

Figure 4.1 Pyramidal organization.

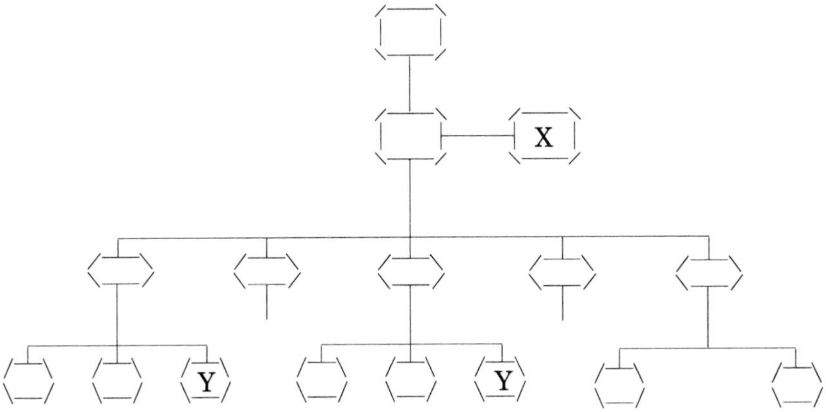

Figure 4.2 Flat organization.

functions (see Y in Fig. 4.2), and the corporate office has a central R&D group located near some basic knowledge center (see X in Fig. 4.2).

In an informal organization, there is some undefined, loose network of friends who scratch each other's backs and pass information through the organization via the grapevine. One who knows the ropes avoids the red tape and knows how to get information when it's not available through channels. I once knew a senior professional who asked the business manager of his group to get him a clock. The business manager refused, saying that research wasn't done by the clock. The senior was angered and I was amused, because he didn't know the ropes. I told him to go back to the business manager, tell him nicely that he needed a clock in his research, that he didn't care which clock, and ask him simply to pick the clock that he felt was best from two purchase requests, one for the clock the professional wanted and one for a clock that was much more expensive. He did this and got his clock. Knowing the ropes helps to satisfy need.

Organize to Facilitate the R&D Effort

The R&D organization

In universities, the general practice is to organize research along lines of scientific or technological disciplines.[51,52] This practice was carried over into industry, and some industrial laboratories still have such a discipline structure or the remnants of it. However, innovation is generally based on a multidisciplinary approach. Some universities have attempted to combine disciplines for greater cross-fertilization and creativity, and most industrial laboratories have some form of product line or functional organization that reflects several disciplines working together.

Industrial problems are multifaceted, and research is multidisciplinary.

Still another R&D organizational structure is the matrix organization in which various disciplines form the permanent departmental structure and various functions form temporary or semipermanent departments (see Fig. 4.3). The functional groups draw on the various disciplines for members, and when a project ends, the members return to their discipline homes. The functional group manager is primarily responsible for managing projects that lead to commercialization, and the discipline manager is primarily responsible for technical competence.

While promoting creativity, the matrix type of organization is hard

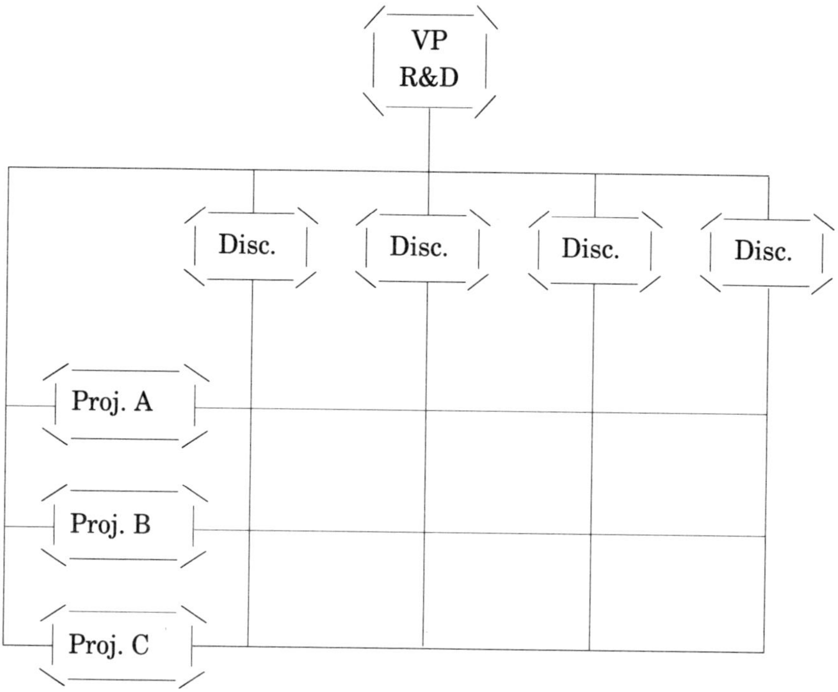

Figure. 4.3 The matrix organization.

to manage, because staff report to two managers. It is not unusual, therefore, to find a mixture of organizational structures, that is, both discipline and functional departments plus task forces similar to the matrix structure.

The use of task forces can bring in other corporate functions and aid in coupling. For example, engineering, marketing, and manufacturing can be members of an R&D task force. These members can participate as part- or full-time project staff and as members of a steering committee. Such teamwork assures appropriate input from other functions and accommodates eventual technology transfers. There is also the committee (probably formed at the corporate office) that cuts across divisional lines. It is a task force that can be temporary or standing. It assesses a problem, asks for certain studies by the divisions represented, and provides recommendations back to corporate and all divisions affected.

Consider the task force for special problems.

Other variations in organizational structure

As noted above, we can have R&D groups in centralized or decentralized organizations. Especially when a company's divisions are geographically dispersed and grouped by product lines, R&D in these decentralized organizations adapts to providing the on-site expertise needed. Although such operational R&D groups may best answer day-to-day problems, the central R&D group is best suited to respond to longer-range problems.

The central R&D group can have many forms.

A central R&D group can have still other organizational structures. For example, it can have a full-fledged complement of disciplines, experts, functional inputs, and support personnel with every facility needed, or it can simply be a group of people with varied experiences and no laboratory whatsoever. Between these extremes is an interesting structure in which a small group of professionals and technicians, say, around 50, form the central R&D group, with additional expertise gained from consultants and contract research groups on an as-needed basis. The philosophy is that such a small group communicates well, can be more creatively productive, and can have the advice of top consultants. As such, this structure can be more cost-effective than other organizational types. Such a group, of course, depends on operational engineering groups for follow-through to commercialization and on the formation of temporary task forces to handle a large R&D project, for which additional personnel would be hired.

Sometimes a new product-development group is organizationally within an engineering group, in which the priority of work is on day-to-day engineering investigations and people or teams are shifted to accommodate engineering needs. Management must protect the new product development group from such manipulation. This doesn't mean that the group should not contribute to solutions of company problems, but it does mean that its major duties are the development of new products and (or) upgrades. It should not be a service group supporting engineering, although engineering may need and have a service group.

A new product group should have a separate identity.

If you must work with such a group, discuss the tasks to be done with the vice president of engineering. He or she can foster collaboration and ensure that ideas from your group get attention or that requests from the engineering group are stopped. If the top person doesn't want to get involved, minimize your own cooperative efforts.

Make sure that requests from other groups have the blessings of their top management.

Another organizational structure is a new product group that takes a product or process idea and carries its development through to a commercial product. Examples of two organizations that use such groups are Rexnord[53] and 3M.[54] Depending on the mission, such organizations overcome technology transfer constraints and the need for coupling and special persons. However, with a broad spectrum of interests, which may also vary over time, the continuing need for a great variety of engineering experience and resources may bring an unbearable overhead problem.

A new product group that carries through to commercialization may be your answer.

The dual ladder

In most laboratories of any size, there are persons interested only in pursuing their science or technology and those interested only in managing a technical team.

Many times a technical person who has tenure or value corresponding to or greater than that of a manager receives less money and benefits than does the manager. The technical person may wish to continue technical pursuits (and would be of greater value to the company in that role) but would like to enjoy equitable remuneration. In answer to this problem, some companies have set up a dual ladder structure, a technical ladder with salaries matched to salaries of the administrative ladder. This system has worked well in various companies, and I refer the reader to a series of articles in *Research Management* for details[55–59] and to an examination of this approach.[60] Advocates of the technical ladder suggest that some experts are attracted to a company with such a policy. A career ladder[54] for entrepreneurs has been started by 3M, and this bears watching.

The dual ladder permits recognition of special professionals.

Some definitions

Appropriate to this discussion are definitions of line and staff, authority and responsibility, accountability, and delegation.

A *line position* is in the direct chain of command that is responsible for achieving the company's financial goals through marketing some product or service. A *staff position* supports the line position by conducting delegated tasks, providing expertise and advice, or just being a gofer. Don't sell these people short! Seek them out and establish a good rapport. They may be people who provide a bridge to your boss or who will be your project champion. They can often give you greater insight into your boss's goals than your boss can, or they can help you open doors in operations that might otherwise be closed.

Know your boss, but know your boss's assistant even better.

Both line and staff are *accountable* for the results of their activities; however, it is easier to measure a line person's accountability by the profit and loss statement of his or her unit. The degree of accountability decreases from top management down as responsibility for both people and dollars decreases.

Authority is the power granted by an organization to require that something be done by others. *Responsibility* is the ownership of an effort's results. You can delegate authority for a task or area of effort to subordinates and make them responsible for results, but you are still ultimately responsible for the quality and quantity of the results obtained. Good managers *delegate* authority to their subordinates and exert every effort to make them successful. And, interestingly, they (the subordinates) should get the credit for successful performance. Such delegation fosters creativity in people and a creative environment in your group. Having to check with a superior before every action reduces a professional to the status of a child. Limited delegation tempts a person to do it first and *then* ask permission to do it—something you don't want.

Authority and responsibility go hand in hand.

You must be very clear about what it is that you are delegating. Is it to take a course of action, make a decision, and carry out that decision? Or is it to take a course of action, check with you on alternative decisions, and then act on the selected decision? Do not make a person responsible for an action without giving that person adequate authority to accomplish the goal.

As the goals of top management filter down through the chain of command (or accountability), they must not get diluted, twisted, or otherwise distorted. The top dog is the top dog. Believe it.

Innovation Starts with Creativity

You need a creative staff

A requirement of the ideation phase is a creative staff that produces an abundance of ideas that are feasible.[61–64] The more ideas generated by your staff, the higher the quality of the ideas eventually selected. And high quality is important, because we are presuming that the superior idea will yield a product or process that has significant competitive advantages. Since performance in the eyes of the customer pertains to market share and market share pertains to profitability, you, as an R&D manager, must be interested in increased profitability through new products. You want ideas that are feasible, that have a chance to yield a marketable product, and that fulfill a corporate goal. To this end you need people on your staff who are creative, who are knowledgeable, and who have the attitude that anything is possible. You do not need people who are scared, defensive, unimaginative, or willing to quit too soon.

Surround yourself with a superb staff, and give them credit for achievements.

Everyone is creative to some degree, and everyone can enhance his or her creativity. The relatively few people who are highly creative usually are right brain dominant,[65–67] are mentally more freewheeling, and have less fear of "imagineering" than people who are left brain dominant, whose talents lie in logic, common sense, and analytical capability. Creative people have an ability to absorb new ideas and combine them in new ways. They are also flexible and do not get enmeshed in rigid thought patterns that block creativity.

Creative people do not fear failure. Courage to the point of fearless abandon is an attribute of creative persons as they launch themselves into a challenge. Contrast that with the investigator or manager who blanches if an experiment's first results are negative, immediately wanting to close out the project, or with members of technical staffs who wish to operate on the safe side of the problem by making only product improvements rather than speculating on new products and thereby risking failure. Such people attempt to protect themselves from all criticism and are never truly innovative.

Failure may be just a new learning experience.

After a failure, a successfully creative person has the ability to keep trying until success is attained.[68,69] The creative person *learns* from failure and comes up with something new to be tried or tested. Thomas Edison regarded unsuccessful experiments as progress. In his discovery of the light bulb, he tried over a thousand experiments before he found a filament that would last any length of time.[68] Only a few people willingly expose themselves to failure in an effort to grow, achieve, and succeed. The chicken-hearted, bolstered with negative results, escape the problem under the guise of saving time and costs.

Apparently, we use only a fraction (1 to 10 percent) of our potential.[68] Creative persons, therefore, extend their creative potential by accepting challenges and taking risks in a considered way, intelligently weighing attendant risk and reward considerations.

Creativity in the industrial world is based mainly on a person's storehouse of knowledge and his or her ability to visualize a potential pattern from a jigsaw of seemingly unrelated parts.[69] In basic research, an investigator either stumbles onto a link to some fundamental truth (serendipity) or correctly hypothesizes that some pathway or combination of elements will uncover such a link. In applied research, serendipity can play a role, but it is more than likely that a new pattern will be derived from a combination of facts in the investigator's storehouse of knowledge. Formal training is, of course, necessary to obtain proficiency in the disciplines of choice. However, beyond a certain point such training has less and less to do with creativity and can be a brake to it by providing all the information necessary to prove that some idea is impossible. The pioneer in atomic theory, Lord Rutherford, insisted that it was impossible to harness nuclear power![70] Beware of people who know so much about so little that they deny the feasibility of anything new, only to find its practicality proved by another company.

Ideas stem from a person's storehouse of knowledge, so stock that storehouse.

On the other hand, an investigator's continuing acquisition of experience and understanding, obtained by working on different company projects and by reading and hearing about technical advances in fields other than his or her own, often leads to significant ideas applicable to the company's needs in fields foreign to the current field of investigation.[62,69–72] This may, in part, explain why most major innovations have come from industries *not* previously engaged in work similar to the innovation. The transistor was developed outside the vacuum tube companies; its impact is well known.

Expand creativity by acquiring knowledge in fields foreign to your discipline.

If top management desires a creative environment, it is important that it hire managers who desire to work in such an environment and that it actively foster creativity and a creative climate.

Creative managers are needed for creative environments.

I think we can agree that creative people are absolutely necessary and given some free time will produce practicable new ideas on assigned projects. However, managers must recognize that many creative people are wonderful "starters" but terrible "finishers." Creative people can get lost in detail or waylaid by other intriguing ideas. It is a challenge to managers to mate such staff with people who are finishers and who do not give the impression that they are taking an idea away from the creative person but merely offering support in the completion of the idea. These finishers have background and experience gained through carrying ideas other than their own to completion. Interestingly, these persons may be technicians with several years of relevant experience or Ph.D.s with a practical bent.

Foster the creative environment

Another requirement of the ideation phase (and of all R&D phases) is a creative environment. Managers (and managements) are responsible for the environment[73] of their R&D groups, and the complexion of the group (or company) will sooner or later be a reflection of the top person of the group (or company)[74] and how that person orchestrates the entity's activities. This may give R&D leaders (or even the CEO) some painful moments should they one day suspect that they have cloned themselves in subordinate managers who are not cutting the mustard.

Top management must, of course, take total responsibility for the creative environment provided to the R&D staff and the kinds of managers that they hire. They are also responsible for establishing incentive programs that motivate and reward. They are responsible for hiring managers who hire staff and other managers, and they must ensure that all managers create an environment conducive to creativity.

To increase their storehouse of knowledge, the R&D technical staff must be voracious readers. As an R&D manager, you must provide inputs (or access to inputs) so that your investigators may constantly add to their knowledge base. Providing for staff to move around in a division or company can also enhance their creativity by adding to their experience and understanding.

Attendance at meetings and conferences may give some cross-fertilization. However, you must see that staff attend meetings that *extend* their knowledge base, in addition to those that merely benefit current business. Too many times investigators restrict themselves to acquiring knowledge closely akin to their current interests.

The staff must extend its knowledge base.

A company can start an annual science and technology conference to bring together its top scientists and engineers for cross-fertilization, to have workshops on subjects of their interest, and to stimulate them with subjects foreign to their current interests. Such conferences can be of great value to a company via exchanges among employees and among employees and external experts, presentations of awards before peers, and dialogues between senior managers and professionals and top corporate and division officers during special functions at the conference.

As part of the support given the staff, managers should train the staff in practical approaches to problem solving and to creating something of commercial value.[75,76] For example, in problem solving *the first step is to define the problem.*[77] The technical staff should ask, "What is the *real* problem?" Analysis may show that the real problem is something quite different from that originally imagined or may show that instead of one problem, there are several problems. Look for simple reasons and fundamental causes. Don't leap at patched up solutions. Often one problem overlies other problems, and you have to solve several in series and in combination. For example, a sonic meter inaccuracy may be the result of the separate designs of the upstream flow-straightener, of the throat, and of the diffuser working in combination.

A problem may be a many-layered thing.

Often symptoms may mislead by pointing to the wrong cause. For example, an investigator may suggest control of an algae problem by an algicide rather than finding that it is the result of a pH imbalance caused by the presence of a sacrificial element put into the system to control a corrosion problem.

Have the analytical measurements been done correctly? I've seen a new product's activity touted because of results obtained from improper analytical techniques. What's going to happen to this product in future use? Are the materials in spec as ordered? Suppliers have been known to make mistakes. Instead of a problem, an investigator may discover a need to be filled; the need must be defined as carefully as the problem was defined if the need is to be matched with the proper tactic.

After defining the problem, the second step in problem solving is to accumulate background material[76,77] both on the problem (or need) and on possible answers (or hunches[78] about possible answers). Bouncing ideas off colleagues (professional staff and other managers) may give some new perspective or enhance the idea at hand.

Define; get background; refer to goals; create.

Third, as manager, you must ensure that the ideas proposed as answers to the problem reflect the goals defined in your strategic plan.

You can't work on every problem posed to you, especially if it takes you away from your program areas.

Fourth, when the number of ideas (or idea enhancements) becomes limited, your staff evaluates the ideas, possibly with additional help and input from colleagues or a consultant. Results at this stage may send your staff back to the drawing board, but whatever constraint turns up, either it can be dealt with through reference to background material or it will point up the need for more information.

A structured approach to problem solving is brainstorming, and managers should see that their staffs use it when ideas evolved are inadequate.[51,78–80] Creativity is, in part, a function of experience. A group, which has more experiences than an individual, offers a greater probability of providing a new idea, a practicable solution, or a solution within a shorter period of time.

Brainstorming is creation through multiple interactions.

If a meeting with colleagues is appropriate, the first session must be carefully planned and well led or it will be a wasted effort. As leader of the session, you must provide pertinent information to attendees before the meeting. During the first part of the meeting, seek many ideas and defer judgments.[80,81] Let ideas digest. As their number tapers off, initiate evaluation and comparison. (A second meeting a week or two later may be especially helpful in that there has been an advantageous gestation period.)

It is your responsibility as manager to promote these procedures, to discuss them with your technical staff, and to lead and participate with your staff in such procedures until they become a way of life. They will give impetus to the ideation phase and to problem solving, saving both time and costs.

Managers promote technical interactions.

An environment that asks for creative behavior (and rewards it) helps close the gap between creative potential and creative productivity. You can enhance creative productivity if you keep in mind the following ideas and encourage your staff to follow the suggested actions:

- Be sure that your technical investigators are interested in problems and have a desire to solve them. A manager's interest helps generate staff interest.
- Background knowledge is, of course, vital. Managers must provide for such information gathering, for seminars in separate yet related fields, and for challenges. Have your staff obtain information on

technical and market trends. Ensure that they are aware of company needs. Require that they set aside time to read in fields other than their own. They should constantly take notes on new information from the literature and on their own ideas.

- Encourage the staff to form the habit, first, of thinking why an idea is good and, second, why it may not be practical. It's all too easy for someone to say why something can't be done rather than to pitch in and help to make it useful. Managers must fear the person who constantly undercuts the creative effort with negative thoughts. Encourage the staff to think, and then to think again.
- Encourage the staff's interest in hobbies that are intellectually challenging such as chess and bridge; such games exercise their creative powers.
- Be sure that the staff understand a problem or an opportunity, read about it, and seek information. Don't let them be discouraged by previous failures to solve the problem.
- Have the staff define the problem. Where are the boundaries? Can they subdivide the problem? Does it overlie other (hidden) problems? Is the staff looking at the real problem? Managers should be wary of accepting the quick answer and should ask for fundamental causes and effects.
- The staff should list approaches to a problem and then look for analogous situations. Synthesize. Let imaginations soar.
- The staff should not be discouraged by failure but should persevere. They should suspend critical thinking for a while if there is no progress; let the problem simmer in the subconscious. Then reexamine it. Take a different tack. Talk with colleagues who may suggest a new approach. Sometimes in talking about a problem, you find you have solved it.
- Research is strewn with failures. The staff must learn that failure constitutes new knowledge that simply must be put into its right niche.
- Carry a notebook for ideas. You never know when lightning may strike!

That you have hired creative people and are promoting a creative atmosphere does not mean that you adopt a laissez-faire environment. As manager, you must direct the staff's creative efforts toward fulfilling corporate goals as stated in the strategic plan. As you will remember, the goal statements contain both a task and a time period within which the task is to be accomplished. Such a directed creative process will have a high probability of producing results, whereas the unplanned or undirected process optimizes the chances of producing the academic and the unsalable. We want, of course, to screen all ideas so that only those of potential promise get into the innovation pipeline.

Effective efforts are directed efforts.

Questions for managers to contemplate:

- Do I have a creative staff?
- Do I need to cull some staff and add others?
- Are the staff getting a balanced diet of information as a basis for ideation?
- Am I doing my best to support my staff?
- Do we (the company) have an appropriate and adequate reward system?
- Is the staff improvement program adequate?
- Do we (the staff and I) practice problem-solving techniques such as brainstorming?
- Do the staff have high performance expectations of themselves and others?
- Are our creative efforts fulfilling corporate goals efficiently?

Project Initiation Requirements

As stated in Chap. 1, the process of innovation usually starts with a specific idea conceived by the R&D group, although the stimulus may have come from other groups such as manufacturing or sales. This idea must satisfy three requirements stemming from various organizational responsibilities before reaching project status:

- It must mesh with the strategic plan.
- It must pass review.
- It must answer feasibility questions.

If the idea does not initiate in the strategic planning session, the first requirement is that it fulfill a need, thereby providing a new opportunity, support an opportunity or answer a threat, and support a corporate goal.

The second requirement is that the idea be worthy of an expenditure of time and funds as determined by the idea review committee.

Therefore, the investigator must prepare for idea evaluation, a more or less formal procedure. To do this, the investigator talks about the idea with colleagues, goes to the literature, makes a computer literature search, checks on the potential market, considers possible performance advantages over competition, and considers what more needs to be done (for example, in the light of the probable technical, legal, and market constraints) to evaluate the idea.

Ideas must fit the mission and goals and be supported by background data.

If an investigator sneaks in a few experiments before having considered relevant background material and determined whether the approach has promise, you can't fault the enthusiasm shown but the person may be wasting time and funds. As manager, you don't want to reinvent the wheel nor do you want to support research that others have found inappropriate for your purposes.

Of course, if the investigator searches the literature and finds relevant experimentation that seems to be erroneous (conditions were wrong, equipment was not adequate in today's terms, or other errors existed), then by all means the literature experiment should be reproduced under conditions considered appropriate. The important issue is that the literature was searched and thought given to the matter before experimentation.

With preparation for idea evaluation accomplished, the investigator sends an idea record form (see Chap. 6) to the idea review committee and obtains an evaluation. The review committee may request that the inventor provide more information, may reject the idea on some specific grounds, or may recommend the idea for a feasibility study. Some idea review methods are discussed in Chap. 8, but any method used must account for uniqueness, markets, company relevance, costs of R&D, probability of success, and so on.

Idea reviews must be constructive.

As ideas are put up for review, you may see in some staff member a fear of criticism. You must moderate this fear immediately. As an R&D manager, you should generate a climate in which criticism is viewed as a constructive activity. Your staff member should be made to feel that if someone's idea is shot down, you know that the person will come back soon with a better one. Fear of criticism (rejection) is not easily changed; it has probably been built up over the years until it has become an intrinsic part of a personality. However, open discussion about idea reviews and the value of criticism, and seeing others accept criticism in a constructive way, will gradually erode this fear.

Also, if it is apparent that an idea has been improved during critical discussion, the evaluation is then recognized as having a positive result. Since the review committee is made up of professional peers, it does not represent the "opposition." After all, the next idea to be reviewed may be one of theirs.

Criticism in the right environment should not be feared.

If the idea review committee accepts an idea as worthy of a feasibility study, the third requirement is that it pass technical, legal, and market feasibility. This is a first pass through constraints in these areas. You or the investigators also discuss the idea with operations to establish their interest, and obtain input from engineering, manufacturing, and marketing.

The feasibility phase will theoretically and experimentally answer the more important constraints in a gross way. You are not looking for detailed specifics at this point but for go or no-go answers. You don't want a formal big-team effort; you want the smallest effort possible to establish a platform from which to judge feasibility. (Peters[82] has published an interesting paper that elaborates on such efforts at this point in the evaluation procedure.)

Sometimes a feasibility study is conducted on an idea that has no company relevance, because the product is in a rapid-growth area of technology and uses an aspect of in-house technology. If solutions to the basic technical constraints have promise and the proprietary position looks good, you should consider how the company might make money. For example, suppose your group excels in R&D for organic coatings useful to resist corrosion in heavy equipment, and a professional conceives an idea for a superior organic conductor based on the properties of one of your materials. Should you promote getting into organic conductors? With no relevance to any company interest, the answer to this one is "absolutely not!" This would be a divergence from company strengths, and the void could not easily be filled in this particular field. However, one way to make money would be to license the proprietary knowledge to a company manufacturing conducting materials, and stick to your materials technology as perhaps embraced in an intermediate product sold to the licensee. With this approach, and with three aspects favorable (solutions to constraints, proprietary basis, and a way to make money), you would submit a report to corporate management and ask if you should proceed to establish your position and, say, to participate in the negotiation of a license arrangement and an agreement to sell the intermediate product.

An idea with no company relevance must have special promise and management OK.

Know that an idea is technically sound and has a reasonable probability of yielding profits.

It is in these early stages of idea or problem evaluation that your research can get started in the best direction possible, based on available information. A random, scattergun approach will not only waste time and funds but also confuse the cleanup crew for some time as they filter out the facts. For optimum progress, determine the feasibility of an idea by:

- Identifying possible constraints
- Obtaining technical, market, and legal background material with which to assess corresponding constraints
- Setting forth and carrying out a minimal study to answer the major constraints in a go or no-go manner

On the basis of proof of feasibility, a manager establishes a project to produce a prototype. The investigator drafts a program for the project based on the constraints that were identified and that will be solved in the development phase.

Establishing and Managing the Project

If the feasibility study is positive and market research supports the idea, the investigator proposes a research and development project based on his or her idea. The feasibility study has given results from which you (the manager) conclude that the identified constraints apparently are not barriers, the possible performance characteristics give important product advantages, and the markets are of reasonable size and character.

Are constraints possible barriers to success?

The investigator should now think in terms of dividing the work to be done into sequential phases. If the investigator accomplishes the first phase favorably, the next phase will be tackled, and so on. This division of work gives milestones to be reached in targeted times and within estimated costs.

Divide the work into milestones.

The investigator defines the first phase of the experimental program in greater detail than the balance of the experimental phases since subsequent phases depend, in part, on the findings of the first phase. In the initial work, the investigator must overcome the constraints in order for further development to be successful. Most projects that have reached this point have extensive analytical efforts associated with them to minimize the more costly experimental phases and reduce the time to market. Although much can be done with the computer, nothing replaces the critical experiments and intelligent interpretation of results.

Assessment of constraints also helps to determine the probability of technical success. The R&D manager can compare the project with others in the portfolio and ask if a project having a certain probability of success and a certain potential for profit is warranted. Management can then compare total risk versus potential benefits. Your assignment of staff to various projects, discussed in the following section, must reflect your knowledge of staff members' individual strengths and weaknesses and your understanding of their needs.

Deploy personnel wisely

The intelligent use of people on projects is important to get the best input, to use to advantage the top people, and to run an efficient cost center. A senior person can handle up to six projects and tasks. I've seen people tilt at the seventh. A good technique is to give each senior person a main project and a couple of second-priority ones, along with some short-term technical service or consulting tasks.

You will, of course, assign projects based on individual capabilities and also try to provide your people with some challenging assignments. I am reminded of my first industrial assignment after my Ph.D. It was a lousy, pedestrian, mundane problem, and I was depressed. I complained to my former major professor, who said, "Englert, there is research there. Find it." He was right, but it would have been helpful had the manager discussed the possible challenges with me.

Put the right person in the right spot.

Challenging the staff is important. I remember asking a manager to read a book on needs for the next decade and tell me if there was anything we should do. He identified a need for a quiet control valve; subsequently he designed a new, simple quiet control valve mechanism. He enjoyed the challenge (and the pleasure of a successful effort).

At another time, I did a technical planning study on refractory materials and wondered whether proprietary products existed among a certain class of alloys and whether we could produce them by practical methods. One of our managers looked at the literature and started a small exploratory project in collaboration with an outstanding university consultant. This eventually led to a reasonably large and valuable program and a new area of interest for the manager. Such assignments can often lead not only to a new product but also to a series of related projects, and to an entirely new career for the investigator. Of course, tasks must be compatible not only with a person's skills but also with his or her interests.

Challenge the staff with new frontiers.

Sometimes a proposal is written by one investigator, but because the person is reassigned by the time the proposal is approved, the project is given to another investigator. This may invite the scorn and disapproval of the second investigator unless some latitude is written into the method of approach and scope of work statements. Such latitude allows project investigators to use their own creativity and become personally involved with the success of a project.

If you are to use the brainpower of your senior personnel in the best possible way, you must provide them with able and cooperative technicians and support people. You must also be aware of the strengths and weaknesses of individual staff members. Combining strengths not only can overcome individual weaknesses but often can lead to a synergistic output. Remember that starters need finishers.

Combine the strengths of staff and anticipate a synergistic output.

Find and use facilitators

If the innovation process had all company functions participating and managers who were gung ho about creating the right working climate, we probably would have no need for special persons. This need arises from the imperfections that inevitably exist. Without a good product manager or project champion, a project has a great probability of failure.[83,84] This unique individual has enthusiasm for new technology and an ability to take considered risks[85] and has a good blend of technical and marketing backgrounds. The champion can identify strongly with a concept or an R&D project and, as the person fostering this concept or project within the company, functions as an entrepreneur.[86] The champion also has political astuteness with a good knowledge of the company and can, therefore, find ways to circumvent formal procedures and channels.

A project champion seeks the shortest route to success.

Langrish,[87] Wolff,[88] and Roberts[89] have discussed the need for special persons. I have also found it useful to the progress of a project to have a person in position of authority (or influence) understand the project and its possible importance to the company to the extent that he or she would champion the project. If this person has the respect of top management and your superiors, your job of selling them on a project will be helped by your champion's enthusiasm for the project.

Cultivate special persons and reap special benefits.

The characteristics of creativity, good communications ability, initiative, and risk taking when exhibited by an above average technical person define the technical entrepreneur—a unique and important person. The correlation between the presence of the entrepreneur and the success of a venture is so well documented[90–92] that it is incomprehensible that many companies do not purposefully look for and use such people and, indeed, establish an entrepreneurial climate.

In 1980, Lehr,[93] then chair and chief executive officer of the 3M Company, attributed the success of 3M to the entrepreneurial spirit fostered by that company. In the preceding five years, 20 to 25 percent of each year's sales was directly due to newly introduced products or services. On the other hand, we have all read of entrepreneurs who started new businesses because they were stifled or looked upon malevolently at their previous place of employment. To get an entrepreneur to move around willingly in a company probably requires consideration of special rewards based on performance. This can produce problems for the manager in obtaining permission to give a special award to the entrepreneur and, if approved, in explaining the reward to other senior professionals. Yet, once done, the practice will be accepted.

Another person who brings people together, who gets information transferred or transmitted effectively, and who aids in product development and problem solving is the *bridge* person.[85] The bridging role *must* be performed by managers and supervisory personnel at all levels. Persons successfully performing the bridge role serve as links in informal communications networks, bringing together people working on separate parts of a problem, and stimulating interaction between functions. Bridge persons are needed throughout the organization to knit it together—within functions, among departments, among divisions, and between headquarters and the field. Such a person working with top management is invaluable and is not added baggage or unneeded overhead.

All managers must play the bridging role.

Use consultants and contract research houses carefully

If you wish to run a cost-efficient yet highly productive group, you will consider the use of outside help. Used wisely, consultants can significantly expedite progress, especially early in the research; however, you can also waste time and money.

You cannot afford just any consultant who professes expertise in the field of interest. Get the best. The person may sound expensive; however, two results accrue: you get what you want, and you get it in a very short time. I have obtained synergistic results by having two consultants with dissimilar backgrounds work on a project.

Use the best consultants: if they are good, they are very very good; if they are bad, they are horrid.

Although consultants thrive on retainers, I have had my fill of thinking about tasks to fill their time when they arrive. I have been most successful in using consultants on an as-needed basis, mutually agreeing ahead of time on their specific task and the amount of time they will spend. At the end of that time, we review what has occurred and arrive at decisions for continuing the task, assigning a new task, or sending the consultant home until the next problem.

Where you can spin your wheels using consultants is in market research. Too many consultants have a broad general background; too few have the specific experience you may need. Convey to a prospective consultant whether you want a quick and dirty job, one phase of a larger study, or an in-depth analysis. Don't give up; find the right one. Again, be willing to pay for the results desired.

Legal consultants are generally controlled by your in-house legal staff but have their place in special situations. For example, a Ph.D. metallurgist who is now a patent attorney will be of more help to a new materials program than an in-house generalist.

To find the best consultant, don't use the yellow pages! Talk with another respected consultant and ask for recommendations. If at a loss, call your local university, ask for the senior professional in your area of interest, and ask for a recommendation. Industrial associations may also be able to assist.

Contract research houses usually profess the capabilities to solve almost any problem that you may bring to them. Few offer novel ideas until stimulated by a problem. Each research house will have some specific areas that are outstanding. Determine which organization excels in your area of interest. Make certain that your problem has the attention of the right person or persons. Too often your problem ends up with people other than those who closed the contract. It is people first and facilities second. Your checklist is:

- Interview the expected principal investigator or investigators before signing, to see if specific experience is there along with enthusiasm and a logical approach.
- You may want to write the name of the proposed investigator or investigators into the contract.
- Do you own the results? Will they resell the work once you are gone?
- Is confidence respected?
- Does the contract research house have adequate facilities for examining the problem, or must it buy some equipment on your contract?
- Check contract details.

If all turns out well and the contract investigator does the work as if he or she worked for you, you will remember the investigator for use in future projects. You will also be more amenable to using others in that organization.

Recognize constraints early

In previous sections, the identification of constraints was brought out as important to assessing the feasibility of an idea, establishing the experimental program of a project, and estimating the probability of technical success. If a project is to fail, you want to deep-six it as early as possible, because the commitment in time and costs goes up dramatically as the project passes through the six-phase innovation process. On the other hand, one must not end a project at the first sign of negative data. How significant are the data? What specifically do they tell you? What are the alternatives? End a project based on sound data which have been well considered.

Technical constraints can shoot down a project at any time. If the staff cannot overcome a technical constraint in a reasonable time, the project should end immediately based on the technical evidence; there is no advantage to a lingering death. Also, as solutions are found during a project, the investigator must be constantly alert to the possibility that new constraints crop up. Discussions of progress with colleagues are often helpful in this regard. One constraint often overlooked is the existence of a test method to simulate field use; if you can't simulate actual use, you probably will never obtain a real product. The best field test is one done with a real user.

A quick death, if warranted, is best.

Market constraints must be considered. The market must be of reasonable size in order for you to afford R&D on a new product. It may be that your company has a floor on anticipated annual sales for a new product. When I was a new manager, I proposed a product with a cinch $1 million annual sales and was surprised to have the proposal rejected because the company's floor limit was about ten times that figure. Interestingly, what may be the floor limit in one division of a company may not be the same limit in another division.

Estimation of market size must account for the segment in which your product is sold; the market can be large, but your product characteristics may fit only a small market segment. Watch out for this! Find out the true market numbers before making estimates. How many competitors are there, and do they enjoy reasonable returns on investments? If there is a dominant supplier, how difficult will it be to take part of its market share? (See Chap. 7 for methods of estimating market share.)

Don't overlook market and legal constraints.

Legal aspects may also harbor constraints. Your computer search may turn up some applicable patents, but if your idea looks good, ask your legal counsel for a patent search. Your counsel will return the search results with an opinion on prior art for the idea. Inspect the disclosures and the claims. Sometimes the language of a claim is so general that it causes problems even though you may think the intentions are limited. Go to your legal counsel for an opinion. (See Chap. 8 for details on patents and licensing.)

Also, during R&D you must look ahead to possible constraints that may arise in downstream operations:*

- Performance specifications easily met using prototype parts may not be so easily met using production parts. Pilot production is essential to unearth such problems and solve them before they reach the customer or before a full plant redesign is needed.
- New manufacturing processes may add troubles if manufacturing personnel are inadequately trained.
- Poor process control leads to in-plant competition for machines and backup of parts in process.
- If the total process is a combination of processes, then the total process should be run continuously in a pilot plant. Study of only batch production from interfacing processes before building the final plant will surely lead to an expensive plant shakedown and probably to some significant redesign.

*Based, in part, on a paper by J. E. Gorgens, "Constraints to Commercialization," 3d Ann. Eng. & Tech. Conf., Dresser Industries, Nov. 13, 1978.

An R&D group that obtains the early participation of engineering and manufacturing has a greater probability of recognizing technical and other constraints peculiar to those operations and of arriving at logical problem-solving conclusions.

Plan ahead for technology transfers

If an R&D project is going well and is to be transferred to the next phase, the baton must be passed smoothly. I consider the actions of the R&D group critical to successful technology transfers. With a new and significantly different product, this means change, and change means trouble and may even mean rejection. Wolff[94] and Foster[95,96] note that when a company adopts a new technology, it must "manage" across a discontinuity; that is, a change in technologies can be disconcerting to everyone. This is obvious with a new technology, but the same thing can also occur with less revolutionary changes. A project transfer has only to change the routine or slightly affect the bottom line to cause problems. And since we need change to meet the needs of tomorrow, we must solve this complex of technology transfer so that the innovation process proceeds easily as it provides new products. If we cannot find adequate solutions to the possible problems of technology transfer, considerable time and money will be wasted in the development stages of R&D and its technical successes either will never reach the market or will be commercial failures.

Have empathy for downstream problems.

Collier[97] makes several points about why the typical operating unit is set up to reject innovation. Among these points are the following:

- Discounted cash flow analyses usually favor the near term.
- Engineering is more comfortable with the reliability and safety of current products than with the risk associated with new products.
- Manufacturing enjoys the efficiency of routine processes, not the variable efficiency of new processes.
- The general manager wants to protect his or her incentive bonus from erosion by new research projects that reduce the bottom line.

In addition, I can add that marketing may be leery of any product that requires new marketing knowledge. However, R&D projects must still withstand discounted cash flow analyses, and the results of R&D must still be transferred to engineering and then to manufacturing and sales.

It is up to management to recognize the possible bottlenecks inherent in technology transfer situations and unite the various groups in a concerted exploitation of new ideas. In so doing, management must recognize the indirect effects of its policies and procedures; those necessary to accomplish one goal may work against another. For example, although discounted cash flow analysis is a profitability measure, it is also a disincentive to new-product research (and to a group obtaining bonuses on sales of current products). Having a standard minimum return on investment that all divisions must meet and requiring that all R&D expenses come from the bottom line with no special treatment will surely lead to a retrenchment in the amount of R&D being done, again to the detriment of the long-term welfare of the company.

Whatever directives are issued to cover technology transfers, they had best be issued by top management. White[98] says that a complete lack of a method for handling technology transfers becomes a liability in the long run, so that the research function is never integrated into the business system and hence fails to contribute significantly to the company's well-being. As a manager in R&D, you have to recognize the clout of downstream operating groups; after all, they pay the bills. Therefore, action taken during the R&D phase has a dominant influence on whether a project eventually succeeds in those downstream areas. Top management actions should provide support, but you should not expect that downstream groups will carry out directives expeditiously and to the letter. Nothing is perfect; nor is intent always realized. You *must* think about the future handling of the project by other organizational units and, by anticipating a unit's needs and the personalities involved, obtain early participation of all role players. Think about and deploy innovation's management circle! It is noteworthy to emphasize that successful management of technical innovation requires an effective integration of R&D, engineering, marketing, and manufacturing.[99]

Top management can help with technology transfer.

Badawy[99] talks of the interfaces, or boundaries, between R&D, engineering, marketing, and manufacturing and the need for "interfacers" to facilitate technology transfer between these groups. I submit that two conditions are necessary for successful technology transfer: (1) the R&D managers must obtain early participation in their work of these other functions, and (2) the managements in innovation's management circle *must want* each innovation to succeed and actively participate in the process. Boundaries exist between these functions because of task-oriented differences and interfacers and managements must attempt to blend these differences as in grafting foreign tissue. But these blends need continual therapy from interfacers and managements or, as in the medical analogy, rejection may occur.

You, as an R&D manager, will show empathy for and identify with the goals and problems of the engineering and manufacturing departments by formulating R&D project tasks that facilitate technology transfers and the ensuing engineering and manufacturing tasks shown in Fig. 4.4. Engineering and manufacturing are brought into an R&D project as early as practical, and marketing is an intimate partner during all phases, advising and providing appropriate data and possible marketing strategies. This will help in transferring not only the technology but also the enthusiasm for a product. It is interesting that more and more people are recognizing this need for early collaboration of R&D with downstream groups.[99–103]

Late participation of others invites opposition.

Also Gomory[104] relates how joint programs by research and product groups foster technology transfer at IBM. Indeed, as the project warrants, you (as manager) might want to form a task force made up of research, development, engineering, marketing, and manufacturing personnel. This does not mean that for every project transferred you have to have some big team effort; however, it does mean that you *must* have available all the ingredients in the quantities necessary. This task force can serve as active members of the project team or as members of a steering committee. Problems associated with joint efforts arise from the additional overhead and the loss of experienced people by the downstream groups. This might be solved partially by

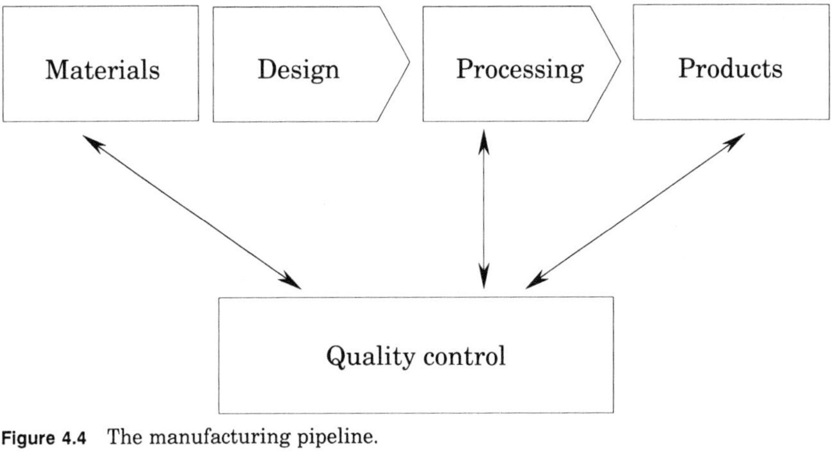

Figure 4.4 The manufacturing pipeline.

some trick financing but it is better solved with an enlightened management. Actions of the steering committee help to answer any reservations of engineering and manufacturing. For example:

- Many new materials may have been considered during the R&D stage and may even have stimulated ideas for the project. Such materials may require a new manufacturing method or may react much differently to conventional machine tools. If so, R&D's early collaboration with engineering and manufacturing is vitally necessary, not only to familiarize them with the new materials and processes but also to gain their valuable insight about practical production methods.
- You must consider materials and design together. Engineering will be particularly helpful in the choice of practical designs. Sometimes a fundamental study of function (for example, a visual flow study) will suggest improved designs using less costly materials.
- There are many methods beginning to be used that are totally unfamiliar to the usual production floor (such as net-shape forming and coatings technologies), and collaboration with engineering and manufacturing is necessary. Training of engineering and manufacturing employees by R&D may be an integral part of the plan for success. Although this example is one of new hardware, the same logic holds true regardless of the specific product or process.
- Engineering can be expected to ask about scaleup of product and processing techniques, so pilot production by R&D will have to be taken to such a size that results are believable to engineering. Product testing by engineering will probably be necessary to convince them of its practicability. The same applies to any continuous operation of several processes; it's better to do this in a small pilot plant than in a large works or plant operation.
- From such R&D operations, engineering will want to know about possible environmental problems, costs of production, and product reliability. Quality control at each stage of processing must be considered rather than just at some final assembly stage. (See the following section on Japanese production techniques.) Performance features and component characteristics that determine performance features must be measured and controlled. Establishment of necessary tolerances to attain performance features is important, but perhaps it is more important to be realistic in such determination. We want to attain specified performance and a product of the best quality at the lowest price.

Lack of good people is often a major constraint in the manufacturing stage. Fortunately, several universities and colleges are offering associate of arts and bachelor of science programs in manufacturing techniques.[105] Obtain a listing of such colleges from the Accreditation Board for Engineering and Technology, Inc.* Such trained persons on the R&D staff will also aid in transferring products. During R&D, managers should ask:

- Will the product answer customer needs?
- Do we have a performance advantage over the competition?
- Do we have the best combination of materials and design according to engineering and manufacturing?
- Are simple quality control methods possible for incoming materials and for each stage of fabrication?
- Do we have a good handle on costs and margins pertaining to what competition can do?
- Will the product be reliable?
- Will the product work as well when scaled up?
- Can we minimize environmental effects of production?
- Does marketing continue to favor the project? If not, what is the problem?
- Do engineering and manufacturing agree fully with the conduct of the project? If not, why not?

It would also be advisable to examine the six barriers to market entry brought out by Porter:[106] Will the economics of scale place a new product at a disadvantage? Can customer loyalty to a current product be overcome? Will capital requirements be a barrier, since your new product carries a greater risk? Will costs the buyer faces in switching to your product be reasonable? Does your company have access to distribution channels appropriate to the product and customer? Is there any government policy which may place undue restrictions on market entry?

As R&D manager, you must also be concerned about the timing of the transfer. Enthusiasm may lead to too early a transfer, and engineering may attempt to design a product before the accomplishment of some required fundamental research. If this happens, they'll find it economically unattractive and abandon the work rather than regroup. I am reminded of widget D transferred from R&D to engineering with great enthusiasm from both sides. Engineering designed a commercial prototype, only to find that they didn't have a crane large enough to lift the main component. Engineering dropped the product. Later,

*Accreditation Board for Engineering and Technology, Inc., 345 East 47th St., New York, NY 10017.

R&D redesigned the product to eliminate metal parts in favor of all organic materials, and it was then in an acceptable state for engineering development.

Choose the time of technology transfer wisely.

When you reach a point where the product functions as it should, the basis of performance is understood (and may even be stated in mathematical terms), and market research still gives a green light, it is time for a total evaluation of the work in preparation for a technology transfer to engineering, where commercial development is the goal. The evaluation starts with a summary of all data, a best design of the product or process as of that date, and an engineering cost analysis. An engineering cost analysis includes the cost of manufacturing, marketing, and distributing the product; estimates of installation costs for the customer; and costs of competitive products. Next, the investigator or the area manager must estimate for the company the return on investment on a discounted cash flow basis over five years and derive an internal rate of return for the customer in comparison with a competitor's product, then estimate characteristics of the market entry and the total market and market share. If all these matters are favorable, technology transfer is feasible. (See Chap. 10 for information on measures of profitability.)

Consider both the consumer and downstream technical needs.

On transfer of the technology to engineering, ideally an engineering manager who has worked with R&D on the project will take charge, senior R&D people will become consultants, and some R&D technical personnel may stay with the project. Having some R&D personnel stay with the project for a while smooths work in the initial period under engineering control.

Unfamiliarity breeds contempt.

To keep communication channels open, always acknowledge the input of other functional groups (or individuals in such groups) to those projects that are winners. Attempt to see that they obtain reasonable rewards for their input, such as companywide recognition of their participation on the project.

Always acknowledge the input of others.

Geographic separation is often mentioned as being counter to facile technology transfer; however, Allen[107] found that beyond a few hundred feet, a few thousand make little difference in the quality of communication. So, whether the separation is down the hall or across the country, the requirements given above appear far more important than the physical separation of groups.

Japanese Production Methods

A favorite question of the last few years is "How do the Japanese produce such high-quality products at such a low cost?" I visited several Japanese companies, and four major reasons became apparent. First, they appear to take much longer to research a project than do their U.S. counterparts. As a result, they know more about a product and are able to get better performance than a U.S. company with a comparable product. For example, I was shown the performance curve for a licensed product, then the higher performance curve for the improved product, and then an even better performance curve for a product that they had changed so much that they considered it their own.

Second, I believe the Japanese are looking at longer-term opportunities than are their U.S. counterparts. Recently a Japanese company ruled out collaboration with a U.S. company because of the U.S. company's short-term orientation.

Third, their devotion to quality control and production efficiency is well known. Less well known is that they base their methods on precepts advanced by such people as Deming, Juran, and Feigenbaum.[108] However, they have improved on such methods and combined just-in-time (JIT) production, total quality control, and participative management to produce the production systems that have by now been viewed by most American companies. Material requirements planning is a sophisticated computer-based system with no consideration given to quality control, while JIT is simple, requires little use of computers, and melds with total quality control. JIT means that parts are made as needed; small quantities are produced *just in time,* rather than large quantities *just in case,* resulting in minimal inventories. Reduction in lot sizes prompts means to reduce setup costs, and the Japanese are most efficient and diligent in reducing setup times from hours to minutes and even seconds.

Their philosophy includes quality control at each stage of production, with workers controlling, monitoring, and improving quality. This reduces the quality control department primarily to general surveillance, monitoring, and training. With quality control at the source, errors are caught and corrected quickly at the point where they occur. Managers encourage ideas from the workers, and workstations sometimes display boards naming the workers and the number of ideas advanced.

The quality circle concept works into means for improving product and process. Quality comes first; output comes second. Although each worker may stop the production line at the sign of a problem, workers from other stations may come in to help in order to keep the line going. Workers are flexible. In controlled automation, the equipment stops when a defect shows up. Visual displays alert workers to a shutdown so that immediate repairs or changes can be made. The intent of such techniques is to minimize labor and the production of defective parts.

By using these techniques, the Japanese have stabilized on high quality, increased productivity, and decreased costs per unit. Such efforts would never have succeeded without management participation, leadership, and support. We can expect JIT and total quality control to be increasingly adopted in the United States in their present forms and in hybrid forms with material requirements planning.

Fourth and last, the Japanese have a good image for servicing their products. They have shown a caring and an efficiency not generally given by U.S. companies.

Summary

- Organize your R&D to meet the needs of the company and your R&D staff; you have several options.
- A task force is a temporary organizational group comprising members that can cut across organizational boundaries, eliminating much red tape.
- The presence of creative personnel and a creative environment is a direct responsibility of the R&D manager.
- To solve problems, define the problem, accumulate background material, don't stray from the strategic plan, and evaluate ideas. In difficult cases, use brainstorming: seek a quantity of ideas among a group of professionals, while deferring judgments. Then evaluate and compare ideas.
- An idea must pass the test of technical, market, and legal feasibility before it can become the subject of a formal project.
- For a high probability of success, a unique, company-related idea must provide more benefits for both the consumer and the company than does a competitive product.
- Placement of the right people on the job, use of facilitators, and appropriate use of consultants are important managerial actions.
- Constraints must be recognized early, answered with some favorable results, and form the basis of project work.

- As the project becomes established, a manager should be planning for eventual technology transfer by collaboration with follow-on groups.
- The Japanese appeared to me to research a product in greater depth and to look longer term than their U.S. counterparts. Their manufacturing methods have led to high quality, increased productivity, and decreased costs per unit.

Cases and Incidents

The Informal Organization

The CEO asks Watts to survey the operating groups working with the R&D group and tell him how everything is proceeding. Watts is to do this in conjunction with Maury from the CEO's staff. Watts quickly learns that it is Maury's idea to do this survey and that he has the respect of the CEO, although they are by no means friends on a social basis. He also has the respect of division personnel.

As they walk into the first division headquarters, Maury says, "Let me talk to Marge, the president's secretary, for a few minutes. I want to get an overview."

Watts is introduced to Marge. Maury asks about Marge's family by name. (Watts later asks Maury if he knows Marge well and is told that he doesn't but that he keeps a file on with all pertinent facts on important people; Marge is the nerve center at this division, and there is not much that goes on that she doesn't know about.)

When Maury asks Marge how things are going, Maury and Watts get a complete rundown on what the president is up to, what his current problems are, who is where in the world, what the R&D problems are, how the president feels about the corporate R&D group's progress, and so on.

At the next division, Maury goes through the same routine but with the division planner and tells him about a planning seminar at corporate. With the chief engineer at another division, he has the same routine and tells him about new equipment at another division and who to contact for further details. Maury repeats this routine at every visit either before or after seeing his contact.

When the trip is over, Watts has not only the formal interviews but also the informal ones on which to draw conclusions for his report to the CEO.

When Watts returns, Shock asks him about the trip. Watts says the trip was very informative. He got a good introduction to some of the company's informal organization through Maury, who also opened some doors for him. Watts said he will have to weed out the straight talk from the gossip, but that won't be hard to do, given the other contacts made. It was a valuable experience. Knowing the informal organization gives one immediate access to the pulse of the company, to information that might otherwise be unavailable through channels, and to a slant on company business that helps formulate one's strategy. Watts says he will give Shock a rundown on the divisions visited after his report is written.

Questions

- What are the formal organizational arrangements in industry?
- What is the informal organization?
- How do you get involved in the informal organization without losing cognizance of protocol?
- Are the politics of the informal organization built on "I'll scratch your back, you scratch mine"?

But I'm Right!

Gerry is transferred from a division R&D group to Shock to supervise pollution efforts in Cory's area. He is a chemist with a national reputation. The corporate public relations office notes Gerry's transfer and makes an appointment for Gerry to give a speech before the local chapter of the pollution society.

When Gerry learns of this, he hits the ceiling and writes a scathing memorandum to the PR office stating in no uncertain terms that it has no right to commit him to anything without prior consultation, that he is busy, and that he has better things to do than to talk to a bunch of environmentalists. An incensed PR employee takes the matter to the CEO, who takes it to Shock, who calls in Gerry.

Gerry is still angry and says emphatically, "Hell, I avoided pollution studies like the plague before coming here. I wouldn't have taken this job without the advancement to supervisor. What do I know at this point about pollution? I have picked up some background in it and know the problems, but I need a few more months before I'll feel comfortable around the experts."

Shock tells him to cool it, that his memo was out of line and embarrassing to the PR department, which knows more about higher politics than they do. Shock says, "We could have solved this in other ways if you had come to me. Now, your memo has hit the fan, and the speech must be given. We will help you all we can."

Later Shock talks to Watts, who agrees that the worst thing you can do is send a blistering memorandum written when you are steamed up about a subject. A memo is so final. It sits and flaps on the receiving person's desk; it begs for an equal and opposite reaction. "Before you write a memo," says Watts, "you have to know what you want to accomplish. In this instance, was it to slap PR's wrist or get out of the speech? If the latter, a memo written on asbestos was not the best way of accomplishing it."

Watts goes on to say that Shock is right that the speech must be given. He also suggests that the subject of memos written in anger might be discussed in a staff meeting. Shock should reiterate his availability for crisis discussion.

Questions

- Could a memo have been written that could have achieved different results?
- Why not discuss the matter by telephone? Who should make the call?

References

51. B. C. Twiss, *Managing Technological Innovation,* Longman, London, 1974.
52. S. P. Blake, *Managing for Responsive Research and Development,* Freeman, San Francisco, 1978.
53. W. R. Stumpe, "Who Pays for New Product Development?" *Res Mgmt.,* September 1978, pp. 17–19.
54. D. Whitefield, "3M: Creative Freedom Is Sustained," Los Angeles Times, Dec. 1, 1985, pt. V, p. 4.
55. D. C. Moore and D. S. Davies, "The Dual Ladder: Establishing and Operating It," *Res. Mgmt.,* July 1977, pp. 14–18.
56. J. J. Smith and T. T. Szabo, "Importance of Flexibility, Job Content, and Individual Temperament," *Res. Mgmt.,* July 1977, pp. 20–23.
57. S. L. Meisel, "The Rungs and Promotion Criteria," *Res. Mgmt.,* July 1977, pp. 24–26.
58. W. D. Emmons, "The Pioneering Research Approach," *Res. Mgmt.,* July 1977, pp. 27–29.
59. E. W. Cantrall, et al., "The Dual Ladder—Success and Failures," *Res. Mgmt.,* July 1977, pp. 30–33.
60. L. Michael-Roth, "A Critical Examination of the Dual Ladder Approach to Career Advancement," (monograph), Center for Research in Career Development, Columbia University, 1986.
61. M. S. Allen, *Morphological Creativity,* Prentice-Hall, Englewood Cliffs, N.J., 1963.
62. J. W. Gardiner, *Self Renewal,* Harper & Row, New York, 1964.
63. C. W. Taylor and F. Barron, *Scientific Creativity: Its Recognition and Development,* Wiley, New York, 1963.
64. R. S. Mansfield and T. V. Busse, *The Psychology of Creativity and Discovery,* Nelson Hall, Chicago, 1981.
65. S. V. Garrett, "Putting Our Whole Brain to Use: A Fresh Look at the Creative Process," *J. Creative Behavior,* **10**(4):239–249, (1976).
66. W. E. Hermann, "The Brain: New Insights," *G. E. Monogram,* March–April 1977, pp. 30–31.
67. ———, "The Creative Brain," *Training and Development J.,* October 1981, pp. 11–16.
68. A. Martell, "Factors for Personal Growth and Achievement," published by A. Martell, Pasadena, Calif., 1968.
69. S. J. Parnes, *Creativity: Unlocking Human Potential,* Creative Education Foundation, State College at Buffalo, D. O. K. Publishers, Buffalo, N.Y., 1972.
70. G. M. Prince, *The Practice of Creativity,* Harper & Row, New York, 1970, p. 101.
71. H. G. Birch, "The Relation of Previous Experience on Insightful Problem Solving," *J. Comp. Psych.* **38**:367–388 (1945).

72. E. A. Von Hippel, "Users as Innovators," *Techn. Rev.*, January 1978, pp. 31–39.
73. J. H. McPherson, "The Creative Organization," LRPS Rep. 426, SRI International, Menlo Park, Calif., December 1979.
74. M. F. Wolff, "Managing the Creative Engineer," *IEEE Spectrum,* August 1977, pp. 52–57.
75. S. J. Parnes, "Do You Really Understand Brainstorming?" in *A Source Book for Creative Thinking,* S. J. Parnes and H. F. Harding, Scribner, New York, 1962.
76. W. G. Hyzer, "First State the Problem," *Industrial Res. Dev.*, September 1978, pp. 143–146.
77. A. M. Biondi, *The Creative Process,* D. O. K. Publishers, Buffalo, N.Y., 1972.
78. J. W. Taylor, *How to Create New Ideas,* Prentice-Hall, Englewood Cliffs, N.J., 1961.
79. W. E. Souder and R. W. Ziegler, "A Review of Creativity and Problem-Solving Techniques," *Res. Mgmt.*, July 1977, pp. 34–42.
80. J. H. McPherson, "Structured Approaches to Creativity," LRPS Rep. 385, SRI International, Menlo Park, Calif., November 1969.
81. S. J. Parnes, "Effects of Extended Effort in Creative Problem Solving," *J. Ed. Psychology,* **52**(3):117–122 (1961).
82. T. J. Peters, "The Mythology of Innovation, or a Skunkworks Tale, Part I," *The Stanford Magazine,* Summer 1983, pp. 13–21.
83. D. A. Schon, "Champions for New Inventions," *Harvard Bus. Rev.*, March–April 1963.
84. "The Role of the Product Champion Conference Board Report," *R&D Mgmt.*, **15**(1):71–72 (1985).
85. J. H. McPherson and D. A. Guidici, "Advances in Innovation Management," *Business Intelligence Program,* Rep. 609, SRI International, Menlo Park, Calif., December 1978.
86. M. A. Moudique, "Entrepreneurs, Champions, and Technological Innovation," *Sloan Mgmt. Rev.*, Winter 1980, pp. 59–76.
87. J. Langrish et al., "Wealth and Failure in Industrial Innovation," University of Sussex, 1972; see also B. C. Twiss, *Managing Technical Innovation,* Longman, London, 1974, p. 16.
88. M. Wolff, "What Do We Really Know about Managing R&D?" *Res. Mgmt.*, November 1978, pp. 6–11 (an interview with E. P. Roberts).
89. E. P. Roberts, "Stimulating Technological Innovation: Organizational Approaches," *Res. Mgmt.*, November 1979, pp. 26–30.
90. D. J. Taffi, *The Entrepreneur: A Corporate Strategy for the 80s,* AMACOM, New York, 1981.
91. H. E. Kierulff, "Finding—and Keeping—Corporate Entrepreneurs," *Bus. Horizons,* February 1979, pp. 6–15.
92. W. E. Souder, "Encouraging Entrepreneurship in the Large Corporations," *Res. Mgmt.*, May 1981, pp. 18–22.
93. L. W. Lehr, "How 3M Develops Entrepreneurial Spirit throughout the Organization," *Mgmt. Rev.*, October 1980, p. 31.
94. M. F. Wolff, "Picking the Right Technology Should Be First Priority," *Res. Mgmt.*, July 1981, pp. 7–8 (an interview with R. N. Foster).
95. R. N. Foster, "Boosting the Payoff from R&D," *Res. Mgmt.*, January 1982, pp. 22–27.
96. ———, "A Call for Vision in Managing Technology," *Bus. Week.*, May 24, 1982, pp. 24–33.
97. D. W. Collier, "The Creative Link between Market and Technology," *Chemtech,* February 1975, pp. 90–93.
98. W. White, "Effective Transfer of Technology from Research to Development," *Res. Mgmt.*, May 1977, pp. 30–34.
99. M. K. Badawy, "Integration: The Fire under Technology Transfer," *Industry Week,* June 19, 1989, pp. 39–77.
100. T. H. Barrett, "Research and Manufacturing Share a Common World," *Res. Mgmt.* March–April 1986, pp. 23–25.

101. H. J. Gray, "Research and Manufacturing Should Be Partners," *Res. Mgmt.* November–December 1985, pp. 6–8.
102. M. F. Wolff, "Bridging the R&D Interface with Manufacturing," *Res. Mgmt.* January–February 1985, pp. 9–11.
103. W. E. Souder and V. Padmanabhan, "Transferring New Technologies from R&D to Manufacturing," *Res Mgmt.,* September–October, 1989, pp. 38–43.
104. R. E. Gomory, "Moving IBM's Technology from Research to Development," *Res. Mgmt.,* November–December 1989, pp. 27–32.
105. A. M. Greene, "Training the Next Generation of Manufacturers," *Iron Age,* Jan. 22, 1982, pp. 62–65.
106. M. E. Porter, *Competitive Strategy,* The Free Press, New York, 1980.
107. E. B. Roberts, "Generating Effective Corporate Innovation," *Techn. Rev.,* October–November, 1977, pp. 27–33; see also T. J. Allen, "Communicating in the Research and Development Laboratory," *Techn. Rev.,* October–November 1967 and T. J. Allen and M. E. Fusfeld, "Design for Communication in the Research and Development Laboratory," *Techn. Rev.,* May 1976.
108. R. J. Schonberger, *Japanese Manufacturing Techniques,* The Free Press, New York, 1982.

Suggested Readings

Ansoff, H. I.: "Corporate Capability for Managing Change," *Bus. Intelligence Progr.,* Res. Rep. 610, SRI International, Menlo Park, Calif., December 1978.

Badawy, M. K.: "How to Prevent Creativity Mismanagement", *Res. Mgmt.,* July–August, 1986, pp. 28–35.

Bailey, R. L.: *Disciplined Creativity for Engineers,* Ann Arbor Science Publishers, Ann Arbor, Mich., 1978.

Buggie, F. D.: "Hurdling the Barrier to Creativity," *Nation's Bus.,* January 1982, pp. 68–69.

Cohen, H. S., S. Keller, and D. Streeter: "The Transfer of Technology from Research to Development," *Res. Mgmt.,* May 1979, pp. 11–17.

Damorell, G. L., C. W. Graves, and G. A. Groenveld: "Technology Transfer in Practice," *Chem. Eng. Progr.,* February 1983, pp. 15–21.

Ford, D., and C. Ryan: "Taking Technology to Market," *Harvard Bus. Rev.,* **59**(2):117–126 (1981).

Gruber, W. H., and D. G. Marquis: *Factors in the Transfer of Technology,* M.I.T. Press, Cambridge, Mass., 1969.

Hayes, R. H.: "Why Japanese Factories Work," *Harvard Bus. Rev.,* July–August 1981, pp. 57–66.

Kottcamp, E. H., and B. M. Rushton: "Stimulating Technological Innovation—Improving the Corporate Environment," *Res. Mgmt.,* November 1979, pp. 19–22.

Moore, A. D.: *Invention, Discovery, and Creativity,* Doubleday, Garden City, N.Y., 1969.

Rubenstein, M. F.: "Problem Solving on Both Sides of the Brain," *Chemtech,* November 1981, pp. 654–657.

Shapero, A.: "Managing Creative Professionals," *Res Mgmt.,* March–April, 1985, pp. 23–28.

Shingo, S.: "Study of Toyota Production System," *Japan Mgmt. Assoc.,* 1981.

Smith, P. L.: "The Seven Principles of Disciplined Creativity," *Across the Board,* October 1987, pp. 43–46.

Stein, M. I.: *Stimulating Creativity,* vol. 1: *Individual Procedures,* 1974; vol. 2: *Group Procedures,* 1975, Academic, New York. Wolff, M. K.: "Hiring People Who do Good Research," *Res. Mgmt.,* January–February 1984, pp. 8–9.

———: "To Innovate Faster, Try the Skunk Works," *Res. Mgmt.,* September–October 1987, pp. 7–8.

Wolff, M. K.: "Hiring People Who Do Good Research," *Res. Mgmt.,* January–February 1984, pp. 8–9.

Chapter 5

Leadership Activities

Questions answered in Chapter 5

- What are the three principal leadership activities?
- What do you do about hiring quickly to meet an urgent need?
- What should you seek to discover during interviews?
- What is your most critical task as a manager?
- Do technical achievers make good managers?
- With what skill do you obtain support from staff and downstream groups?
- How do you motivate people of varying needs toward greater achievement?

Managers have authority granted them by the organization. They may also be leaders if their groups accept their leadership because of their exceptional characteristics and skills. Managers need such leadership qualities in order to rally their staffs into strong, creative working units. Hence, managers who are also good leaders owe their authoritative power to an organizational grant and their leadership power to acceptance of their qualities by their groups.

Leadership activities relate to everything done by the head of R&D and his or her managers to obtain, retain, inspire, and improve staff. Principal among these activities are those relating to the hiring, training, and deployment of personnel; communicating to promote cooperation; and motivating to enhance productivity.

Personnel Activities

Superiors measure managers in part by the results obtained by their staffs. This in turn reflects the quality of staff that managers attract.

Proper recruiting, training, and appraisal are basic to obtaining the best possible staff. Have you ever seen on TV several dogs with their owners, each dog and owner having a close resemblance? Laughable, right? It's not so laughable when the same thing occurs between a leader and a group; therefore, if results are not as desired, look not only at the staff but also at yourself and your style of management.

Managers are measured in part by the results of their staffs.

Obtain and retain only the best staff

No approach to managing people or encouraging creativity will avail if you do not have a competent staff. Whether you are the CEO or a first-level supervisor, it is imperative that you build your group or organization with the best people that you can attract. The successful manager gets things done through and by other people. The manager who has an R&D staff that projects an image of an aggressive, creative, and cooperative group will find it easier to influence other functions (such as engineering and manufacturing) to participate in, and to assist with, the total innovation process.

An effective manager will see that training is given to make a staff member more productive and to allow the person to reach such personal goals as added technical capability or a future supervisory or managerial position. Such training includes on-the-job suggestions, attendance at selected seminars and short courses, attendance at management training courses, review of selected books, and study of audio courses. Discussion between superiors and staff members about their progress and possible future may lead to useful suggestions and ways to support them. A professional's work in different areas of the company and on varying work assignments will lead to an expanded background, which is important to creativity.

Offer training to improve, but also to retain, professionals.

Appraisal for advancement must include consideration of a person's benefit to the company and his or her interpersonal relationships. In an appraisal meeting, I prefer to ask questions and let employees discuss their progress. Employees usually will be quite candid about their work and personal strengths and weaknesses. If not, leading questions may elicit comments that permit assessment of how they are doing and of any problems. (See "Personnel Review" in Chap. 6.)

Let professionals appraise themselves; you ask questions.

There are occasions when R&D staff will not recognize or will not acknowledge such problems as alcoholism, lack of creativity, or lack of necessary skills for the position. You must not procrastinate about discussing serious problems; face them delicately but firmly, straightforwardly, and fairly. In some instances, it is wise to have a second party present, perhaps someone from personnel.

At the end of the review, summarize with the employee what has been said and set goals for the future. I always ask the employee to make suggestions about how to improve the group, and I always give each suggestion serious consideration. In all instances, note the results of these meetings and place them in the person's file.

A plot of managerial or supervisory capabilities versus technical progress gives a graph of a person's direction over time. You first score technical capabilities—creativity, initiative, knowledge, ability to reach goals, what one does to improve technical skills, and so forth—then score supervisory or management capabilities—caring attitude about people, personality, desire to teach or lead others, communications abilities, and so forth. You then plot these scores over time on a chart of technical versus management capabilities and you can see in what direction this person should be heading. As manager you should give support and guidance to what is indicated.

About potential or existing staff:

- There are those who plod along doing an adequate and meticulous job but will never create anything of significance. There are places for these people, and each manager must decide whether and how to use them.
- There are the very creative and personable professionals who believe other people can never create as good an idea as they can. Don't lose these people but don't offer them a supervisory position. Despite capability and charm, such people will dampen the motivation of people under them.
- Then there is the creative optimist who finds it hard to complete a project or who gets so immersed in one project that deadlines or other projects are forgotten. Such persons need administrative or supervisory help.
- And then you find those people who fit in well and are creative but who need continual motivation. With them, you must regularly call their attention to needs and opportunities that are important to the strategic plan.
- Watch your entrepreneurs. Their needs may be quite different from those of others on your staff. They will probably not want to participate in discussions other than how they can best be successful on their own tasks. That isn't all bad, but often they also want to be managers, and then you must cope with their probable lack of abil-

ity to manage. It is best to place such persons in temporary management positions, such as leaders of special task forces, each instance giving them a chance for visibility and achievement.

- Watch for the chronic complainer who can't find anything right and loudly proclaims what should be done; you don't need such people, regardless of their capability.
- In the same vein, get rid of the low performers and the incompetents, but do this astutely. For example, when established goals are continually not met by such persons, you probably have substantial grounds for dismissal, but remember that when you are establishing goals with a staff member, *the staff member* should suggest when and how these goals are to be met; then if goals are not met, staff cannot claim they were unachievable in the first place. Everyone deserves an opportunity to improve, even if such chances are considered minimal. Tactfully point out a person's shortcomings, and give a time frame for improvement. Again, document such meeting discussions and place them in the person's file.

Technical skills do not necessarily translate to management skills

Most high achievers have technical skills that are excellent for their job; they are creative and they communicate well. High achievers on the professional staff will have greater technical skills than their area manager, the area manager will have higher technical skills than the head of R&D, and so on. However, any manager must have technical skills that permit adequate discussion of technical situations. Where the professional staff is creative about technical matters, the area manager is creative about ways to increase staff performance. Where a staff member's oral and written communications present and interpret technical results and state progress toward a technical goal, the head of R&D communicates the same results with a view to meeting the marketing goals of the company. However, a manager must keep reasonably abreast of technical events through technical planning studies, seminars, courses at local universities, and various readings. For example, no manager can afford not to be knowledgeable about computers and computer programming.

Managers get satisfaction from the achievements of others.

So, we have a skill spectrum depending on whether one is in the professional or managerial ranks and, if in the latter, on a person's place in the management hierarchy. The same is true of a person's desire for accomplishment. At the professional staff level, there is a strong drive for achievement (technical accomplishment) and little drive for power (influencing people); at the CEO level, the reverse is true. In the middle of this spectrum, we can have problems.

Faced with the question of becoming a manager, the professional must ask whether major satisfaction comes from influencing people or from achieving results through technical expertise. One characteristic of good managers is that they get satisfaction from what others accomplish under their direction rather than from what they may accomplish in technical research. Thus, senior persons with a strong personal desire for achievement should choose to proceed up the technical ladder rather than the managerial ladder. Entrepreneurs probably will not make good managers because of their single-minded drive. If such persons choose management, they may provide good output until administrative and interpersonal problems get out of hand and then they will lose control of their staffs.

The selection of the managerial or technical ladder is probably permanent, so it had best be a good decision.

There is the instance of the very creative professional whose ideas were always the best ideas; no one else's ideas ever measured up in his eyes. He was very personable but could be cutting in his criticisms. Put such people on the technical ladder so that you don't lose their creativity, but never let them supervise others, for they will kill any creativity in their staff. Another instance concerns a brilliant professional whose ego forced him to be number one at the expense of all other colleagues. He was also pushed by his family's zeal for his advancement. They played the game of buttering up superiors while treating peers in a condescending manner. Try such people in entrepreneurial, never in managerial, activities, for they do not have leadership qualities.

Technical expertise is packaged in several ways.

Just as we generally do not want entrepreneurs in permanent managerial positions, neither do we want managers or professionals in positions not commensurate or compatible with their capabilities and personal goals. "Know thyself" is not an idle admonition.

Always match talents with the job.

Recruit the match you want or don't hire

In several of the preceding chapters, I have discussed obtaining the right people and using them wisely. In this section, I discuss how to get them.

Recruiting is an R&D manager's most important task. It is difficult and it is risky. Because you can't judge the merits of a person on the basis of looks, and because résumés leave much to be desired, other approaches are required. The following may be helpful.

When recruiting, keep in mind not only the precise needs of the vacancy to be filled but also the complementarity of the applicant's talents with those of other staff members.

With people fresh out of school who are being considered for the technical staff, you can get some idea of their qualifications from their professors, and sometimes coworkers can gauge their abilities.

Always verify degrees. Routine checks by the personnel department may miss a candidate's actual status. Newspaper stories of persons reaching high positions without having the degrees they claim are not uncommon, since the reason the stories reached the newspapers was that some catastrophe occurred owing to the person's lack of basic knowledge in his or her field. I can tell you from firsthand experience that waking up to 1-inch headlines announcing the arrest of a senior professional on your staff for acting illegally in another technical field is most disconcerting. It can ruin your whole day. Don't let this happen to you!

Poor recruiting gives black marks to everyone.

You can also get some fix on candidates' sensitivity, curiosity, and analytical and synthesis capabilities by presenting a problem in an informal (and even offhand) but well-considered manner. For example, give candidates one or two of your group's current problems as an example of what they might have to contend with, and see what they say and how their mental processes work.

Another method is to let candidates talk to some of the staff. I am always interested in the types of questions candidates ask; such questions are good indicators. For example, I distrust the person who asks administrative questions of the staff rather than technical questions that show curiosity about the work at hand. I also distrust the candidate who is author of a zillion papers but has no patents. Patents are a good measure of curiosity that had follow-through. One point of caution: Some companies have had such poor experience with the validity of patents and (or) infringements that they have decided to maintain a position through trade secrets (see Chap. 8). Determine if a candidate was in such a situation.

Encourage candidates to talk; they'll either win you over or put their foot in their mouths.

Interviewers should ask questions that:[109,110]

- Elicit candidates' attitudes toward life and R&D.

  Are they positive?
  Are they sensitive to problems?
  Are they curious?

- Establish candidates' initiative.

  Are they self-starters?
  Do they have a strong desire to achieve?
  Are they persistent?

- Determine candidates' dependability and maturity.

  Do they give straightforward, logical answers?
  Are they decisive?
  Are they loners?
  Are their answers considered, or do they give snap judgments?

- Reveal candidates' work records.

  Have they been treated unfairly in any past jobs?
  What criticisms and compliments have they received?
  Has an assignment ever bored them?
  Have they ever failed on an assignment? If so, how did they handle it?
  What do they feel are their strengths and weaknesses?
  What would they eventually like to do?

- Identify candidates' interests.

  Are their interests broad or parochial? (If broad, a candidate is probably flexible, interested in a variety of experiences, and has a greater probability of being creative.)
  Do they have interests outside of work?
  Do they participate in community activities?

- Determine whether candidates' talents are specific to the position.

  Are technical talents adequate or above average?
  Is the candidate creative?
  Is there enthusiasm about the position?

Questions to elicit such information should be well thought out ahead of time. Always check candidates' work history and explore with them any possible concerns you may have, based on their résumés. This may require some reading between the lines and subsequent telephone calls to references.

Let candidates give a seminar. This is an excellent technique to identify the depth of their claimed skills.

Candidates for managerial positions will have had experience supervising groups similar to that needed in the vacant position. They will have read texts and attended seminars and courses on management. Determine their achievement versus power needs, sensitivity to others, maturity, and common sense. Do they have a balance of characteristics, skills, and management approaches to make effective managers? (See Chap. 2.) With a manager (or senior) candidate, get acquainted with the spouse. You can depend on it, a flaky spouse will be paired with a poor candidate.

Communication Activities

You can't do it alone. Your staff can't do it alone. You need to have a united staff that supports and is supported by other company functions. Your communications skills make this possible.

Communications link all activities.

There must be both horizontal and vertical communication. Communication must exist between the head of R&D and his or her staff, between the head of R&D and his or her superiors and their staff, and between the R&D group and operations. Innovation's management circle again!

I recommend a weekly meeting of research managers and the head of R&D. Area managers must keep the head of R&D well informed, since he or she represents them to top management. Each manager should submit a one- to two-page summary of goals, accomplishments, and problems for the past week and goals for the current week. Such mini reports can be the basis of presentations during the weekly meetings. Each manager should sum up progress versus planned tasks, identify needs or constraints, identify decisions needed, and state actions proposed. After a discussion, the head of R&D makes or approves decisions and with associated managers decides on needed actions. Meeting discussions should include administrative matters pertaining to each area and any pronouncements by corporate offices affecting the welfare of the staff.

Use much the same agenda in a project meeting. During a project meeting, each professional should discuss his or her projects and then turn in a one-page description with such items briefly noted. The manager then uses these pages in reporting and project monitoring tasks.

A general monthly meeting of all the staff often has benefits. Highlights, including problems, of each research area are discussed. Administrative matters are aired, and any matter that affects the welfare of the group is brought out. This is a good time to acknowledge new ideas of staff members, present special awards, and compliment deserving individuals.

A staff meeting should be ordered informality.

Communication is a two-way street. Talking is not necessarily communication. We must convey our thoughts in language that can be understood. Answer those quizzical looks showing reservations. On the other hand, it is important that as listeners we "receive" on the same wavelength as that "sent" by the speaker; that is, we must try to understand another's viewpoint. Dialogue is thus fostered. If you are figuring out your rebuttal to someone's argument while the person is talking, you are not listening with empathy. If you are judging, you are not listening. If you are thinking about how you can advise a person, you are not listening. See also Boyd.[111]

The head of R&D must keep management informed about progress and problems. A monthly report should be brief; a quarterly report should provide more information, including summaries of recently acquired technical and market data. Inclusion of a one- to two-page description of an emerging technology and its possible impact on the company can strike special interest. The annual report substitutes for the final quarterly report. In this report, the head of R&D should review the strategic plan, indicate any revisions to the operational plan, and present his or her managers' up-to-date information on technical and market trends.

On a selective basis, the vice president of R&D sends the monthly and quarterly reports of the R&D group to operations to communicate progress of a specific program. Sometimes, portions of quarterly and annual reports can be separated and only those parts that pertain to technical and market trends and briefs on emerging technologies sent to operations. These can spark the imagination of someone in a division and bring the two groups into closer contact.

Visits to operations by area managers and the head of R&D are useful to gain an idea of operations' needs, to communicate what is going on in the R&D group, and to discuss how their results might be used. One of the best techniques that I have used in getting to know people in operations better and in influencing them with regard to an R&D group's capabilities has been to hold 1½-day workshops on specific subjects. During the first day, people from the R&D group, from operations, and from the consulting community talk on different aspects of a selected subject. Open discussions are encouraged. During the next half day, the discussion (with outsiders now excluded) centers on how to use the information presented, concluding with recommended actions. One of the R&D professionals chairs the workshop and provides written summaries to all participants.

Open communications with operations; it will be a beneficial two-way street.

Motivation Activities

The last of the leadership activities to be discussed is how you, as a manager, can motivate your staff to greater accomplishment.[112] To have a staff that reaches its full potential requires not only that you support and motivate them but also that your company have policies which support you in this effort. Most people do not reach their full potential, because they are lazy or because their ambitions are misplaced. It has always graveled me to see a brilliant person not reach his or her full potential.

As a unique human influencing other unique humans, you must consider management philosophies and decide on the model that best fits you and your circumstances. In the following, I present brief descriptions of major philosophies, enlarging on them when I think it is to your benefit, and provide references for your further reading.

Motivation theories

Various motivation theories are listed below with references for your consideration. Of these, the works of Porter and Lawler[113,114] and Cherrington, et al.[115] warrant your special consideration.

Theories focusing on individual needs and drives. Theories that identify the needs that drive people to act include Maslow's[116] hierarchy of needs, Hertzberg's satisfiers and dissatisfiers,[34] and McClelland's power, affiliation, and achievement needs.[27] Maslow, along with Maltz[117] and Martell,[68] believes that people continually seek satisfaction of certain needs and that this search determines their behavior. Maslow identifies these needs as physiological needs (the need for survival and companionship), a need for safety and security (the need to meet society's demands adequately and to develop financial independence), a need for approval from others (the need to belong), a need for self-esteem and self-approval (the need to achieve some goal under risk of failure and disapproval), and a need for self-actualization (the need to fulfill their potential).

Managers who are aware of these needs can use those not yet met as motivators of the technical staff. But remember that motivation must be on a personal basis. One person's need is not necessarily another's.

Drucker[118] says that a need may change in the process of being satisfied. In instances where security needs are reasonably satisfied, economic rewards may still be important motivators. Also, as an incentive reward decreases, its capacity to create dissatisfaction increases rapidly.

Know the needs of your staff.

The middle 1980s were an interesting time in which benefits and pay could be reduced regularly rather than raised regularly. Such situations reverse but can be real challenges for managers, who must lend some logic to such reductions while attempting to retain their top performers (see the following section on reward for top performers). Managers (who are also losing benefits) may have to hire new staff who have no corporate memory and who think of their work as a job rather than a career. As a result, the new staff may work for themselves with little or no loyalty to the company. Management skills are at a new premium!

Theories focusing on why people make certain choices. Expectancy theory states that people choose behaviors that they see will lead to desirable outcomes, such as pay and recognition, that satisfy their needs. Equity theory states that people want a reward-to-effort ratio similar to what other people get doing the same work. The Porter and Lawler thesis[113,114] states that people will exert effort after considering the possible reward and the probability of getting it, and when a reward is given for differential performance and perceived as equitable, satisfaction follows. Greene[119] concludes that the Porter-Lawler thesis can be applied to merit pay if the manager can differentiate among performance levels. As a basis for a theory on performance-contingent rewards, Cherrington, et al.[115] found that increased performance was obtained from rewarded high performers and nonrewarded low performers. On the other hand, a nonrewarded high performer's later performance dropped and a rewarded low performer's later performance stayed at the low level. This is in line with Porter and Lawler's thesis that you reward the superior performers.

Pave the way for achievers to perform; then reward performance.

Concluding remarks on motivation. As I've said before, *you must get to know the people on your staff as individuals if you expect to influence them.* Then you can match the type of motivation with a person's particular needs at a particular time. We must also account for the characteristics of R&D professionals[18,41,120,121] as discussed in Chap. 2.

You motivate a person by rewarding above-average performance, by giving general encouragement, by assigning challenging projects, by providing a stimulating environment, and by providing the support needed for accomplishment. Offer the staff rewards for performance, and if they succeed, give them the rewards. You don't reward to get happy, satisfied people and then expect them to produce. High performers must take considered risks; the environment of your group should encourage such actions and label failure as a learning event. However, you don't reward someone for a project well done that ends in failure. Reward success.

I submit that some pressure to produce is good; I've found that when the pressure starts to rise, the high performers will produce to lessen the pressure. Margerison[122] says that if hygiene factors such as pay are satisfactory, managers who support and have high expectations for their staffs can expect the staffs to fulfill their expectations. Manners[123] says that the capacity to motivate depends on a manager's credibility (discussed earlier).

High performers work to lessen pressure; low performers aren't aware of it.

The staff will be interested in the equity and fairness of their treatment by managers. Consistency of action is paramount, yet in the case of rewards, company policies and procedures may not allow the degree of freedom a manager may like in attempts to be consistent. In such instances, managers must improvise and also exert every effort to improve the situation.

Pay and incentive programs

Your corporation must have adequate pay and incentive plans to have a credible image as a top innovator. If the program is already a good one, wonderful; if it lacks something, it is your duty as head of R&D to shape the program with help from the personnel department. You should get acquainted with programs of competing firms and of selected distinguished firms in the nation. After deciding what you need, present a well thought out plan to your superior, *placing the plan in the perspective of increasing productivity.*

Seek an equitable pay program, or lose your top professionals.

Any remuneration program intended to attract new personnel and keep valued professionals is of limited value if such basic expectations as job security, rewarding job content, freedom to act independently, and so forth are not also evident. Except for the special project for which professionals are hired with the prior knowledge that continued employment depends on the success of the project, you hire for a long-term relationship. Both sides must take due consideration of their compatibility should they enter into this relationship. If you are fortunate, your company has no history of repeated or cyclical layoffs which indicate inherent instability and job security is not a question. Job content should provide personally rewarding challenges and experiences. Also rewarding is the entrepreneurial spirit of the group, the freedom to use one's innate creative abilities, and the possibility of independent action within the house rules, that is, freedom to create within the group's mission and the research areas laid out in the strategic technical plan. The wise company, of course, will make continuing training and development available to avoid technical obsolescence and to expand staff abilities.

Instability in pay or incentive programs will hurt recruitment.

Pay program. In addition to its position on professionals' basic expectations, the company must have a sound pay program. Salaries must have equity with local markets. This may cause some problems on transfer of personnel, but that is less important than local market equity because the company must compete on the local market or lose personnel to it. Managers must continually lobby with the personnel department to maintain an equitable pay program.

Salary increases are based on either performance ratings or adjustments to the market. When the company limits the pool of funds for raises, it may be difficult to reward significant performance without adequately concomitant penalty to someone who performed adequately but less significantly. What happens then is a narrower spread of percentage increases than the spread of performance would suggest, or the high performers are rewarded and the adequate but lower performers are not rewarded proportionately, if at all. Porter and Lawler recommend rewarding the high performers adequately; I believe actions depend on the total remuneration package and the circumstances.

Reward the high performers!

The manager must be concerned not only with market equity but also with equity among staff members. During an appraisal, there will always be people who express disappointment with their raise and argue strongly for adjustment. Consider the arguments carefully and take appropriate actions, including providing a carefully considered explanation to such staff members. During appraisals, make certain that you become aware of every staff member's personal concern for security within the company. This may be the only time that you get the staff's true thoughts about such matters, although an open door policy should welcome discussions of such thoughts at any time. Enter a summary of the appraisal meeting into each person's personnel file.

Make appraisal times profitable but not the only times staff can seek a dialogue.

Promotions can be used to recognize significant progress or to adjust pay upward. I don't like to resort to promotions for salary adjustments, but there have been instances where the salary schedule would not allow equity by any other means.

Merit raises usually go along with annual appraisals. There are times, however, when raises must be made at other times in order to adjust to the market. Also, as one attains greater seniority, times between merit raises may lengthen or if combined with the annual appraisal, the normal merit percentage may decrease. Merit raises should contain an element of adjustment for inflation and an element based on performance. Performance should be judged primarily on successful results and secondarily on factors such as self-improvement and personal relationships.

Incentive program. The incentive program offers additional rewards contingent on personal performance and can be designed to include several types of incentives.[18,124,125]

Advancement up a technical[55] or entrepreneurial[54] ladder can be made as a reward or an incentive.

A small monetary reward usually is given on filing a patent application, and a certificate on patent issuance. Managers should rigidly review the filing of patent applications to avoid having professionals paper their office walls with certificates having questionable merit.

Special awards encourage creativity, innovation, and cost reduction. These are separate and distinct awards requiring different criteria for evaluation. Medals, certificates, dinners among peers, and recognition in the company newspaper are common; money or shares of stock are also often given. Such awards would start at some organizational level within the company consistent with the type of award. Cost reduction awards might start at some grass roots level, and creativity awards would start at the division level. There would be selections of individuals and their awards at each sequentially higher organizational level until selection and awards would be made at the corporate level. Such awards can be made very timely within operations but only annually at the corporate level. Sometimes, a vacation or a large sum of money is given on a one-time basis as an award for a significant accomplishment.

Make awards meaningful, hard to get, and prestigious.

R&D groups within operating divisions generally share in the division bonus plans, but corporate R&D bonus plans vary widely, and many companies have none.[124–128] The general objection stated for corporate R&D bonus plans is that no mathematical basis exists comparable to the existence of profits within operating divisions. There are also those who believe that the R&D person has reward enough given adequate pay, a chance to do significant work, and recognition by peers. As one superior put it, "Englert, you should pay me to work here." (Hardly a positive motivational stimulus.)

I suggest that goal setting can be used in establishing the basis of an incentive bonus plan in an R&D group, with some minimum set of goals expected to be met and performance beyond that minimum rewarded. These goals would be based on the success achieved in R&D and on self-improvement, enhanced personal relationships, and so on, as in the merit raise. Commercial successes arising from a person's R&D could also be factored in, although these are generally several years distant from the person's main contribution.

Performance beyond set goals is one criterion for an R&D reward.

Since achievements are generally a team effort, the mathematics of arriving at bonus distribution could include some points for the senior investigator, some for the team, and some for the manager of the team. Awards would vary from 0 to 50 percent of base salary, depending on an assessment of differential performance. A team of technical and nontechnical members of the company (preferably top management) should advise the head of R&D on amounts and distribution among the staff and managers. A pool of funds must be set aside by corporate for this purpose, and this may moderate the advisory team's recommendations. See also Mower and Wileman.[129]

It is important that such awards never become automatic; no one should expect to receive a certain percentage of salary every year as a bonus. In the same context, bonuses should never be given with the intent of giving a lower salary than the person would otherwise obtain in the marketplace.

Incentive awards are not a substitute for salary.

Managers should look at other awards to see if any fit their particular situations. An amount of free time should be given creative professionals, say, 10 to 20 percent, in which they can explore some pet ideas. Sending staff members to special meetings or to special technical or management courses can be real morale boosters. A luncheon honoring an individual's achievement will have a good affect on both the person and colleagues. Neither does it hurt to recognize achievement in staff meetings and the company newspaper.

An award, monetary or other, is recognition for exceptional achievement, and recognition is something for which everyone hungers.

Summary

- Principal leadership activities concern recruiting and training personnel, promoting communications to obtain cooperation, and motivating to enhance productivity.
- Obtain the best staff you can; the company's and your welfare depend on it. Recruit the right match for your needs, or don't hire.
- Interview candidates to determine attitudes, self-activation, stability, work record, interests, and talents. *Recruiting may be the manager's most important task.*
- Excellent technical skills do not translate into excellent managerial skills.
- Use your ability to communicate and your management style to influence and gain the cooperation of everyone in your management circle.
- Know your staff as individuals if you expect to motivate them; match motivation to a person's need at a particular time, since needs vary with each individual and circumstance.
- Treat your staff with dignity and equity; seek the best possible rewards for them; expect goals to be accomplished and personal potentials to be reached.

Cases and Incidents

Hiring Under Pressure

Shock tells Watts he pulled a real boner in hiring Smith, the head of the engineering services area. He gives Watts a brief rundown on what has happened.

Smith has a degree from a university not well known for its engineering efforts, but references indicated that he had a reasonably broad background and showed initiative in prior work. A former client when Smith was consulting testified that Smith had solved their problems in a reasonable time. One of the R&D staff's professionals (not an engineer) interviewed Smith and gave him reasonable marks for his interview.

"Remember," says Shock, "this was also a period when engineering personnel were scarce and a time when our need was urgent, so we hired him. But now Smith found himself in an R&D group of a large corporation with a great variety of products and services. His interaction with the other areas in the group was not positive, and he wasn't able to meet their engineering needs.

"It is not that he didn't want to help," Shock continues, "but he just couldn't seem to give quick guesses on probable answers or provide good ideas in the time we could afford, and that's what we need. Harry, Smith's senior engineer, was also getting more and more critical of Smith's ability to respond to the needs of the group."

Shock looks suddenly tired as he continues. "Smith was willing, but either he was just not trained well enough or his experience was not broad enough to provide the input needed. I decided to let him go. In private, we agreed that another chance wouldn't lead to a different conclusion. So, he has been given three months to conclude his current work and to find new employment. We have given him some names of candidate employers and let him use his office and the group's secretary before he leaves."

Watts shakes his head before commenting. "This is a fairly typical story. Your group was under pressure to produce; good people are hard to find; pressure builds; group hires warm body without adequate checks. You must seek a first-rate staff in order to reach company goals quickly and creatively. A candidate must satisfy the needs of the group and mesh with the group culture. However, it's best not to hire at all if you can't find a candidate matching your criteria. Farm the work out or turn it down."

Watts looks sharply at Shock, "I hope that you placed everything discussed into Smith's personnel file!"

Questions

- What would you have done? Not hire Smith?
- How could the initial interviews have turned up Smith's inadequacies for this position?
- Don't you bring top-level people to the company for an interview so that both parties can be satisfied?
- When hiring, when do you use an outside agency? In what and where do you advertise?

If There's No Recognition of a Problem, There's No Solution

Dr. Chen, manager of the organic coatings area, asks for a private meeting with Shock. His problem is Charley, a new polymer chemist with good credentials.

Chen says Charley's a damned drunk. He comes in usually on time except when he's loaded; then he's always late. He smells of gin, and sometimes Chen thinks his eyes are going to bleed all over the floor. And his lunch hours are suspect.

Shock asks if Charley's work has suffered. Chen replies that it's not too bad, but he's afraid that Charley may do something accidentally that will foul up some experiment. After all, they do work with chemicals that are potentially dangerous.

Shock asks if Charley's activities have been affected in any way. Chen responds that the most obvious effect is his appearance and slurred speech; because of these Chen can't have him interface with visitors, and that limits his usefulness. Larry has also complained about Charley's behavior.

Shock says he'll have a talk with Charley and see what he has to say. Chen leaves.

Shock calls Charley in for a conference and says he has seen a change in him since he started about six months ago. He'd like to see if he can determine the cause and maybe help get the old Charley back. Shock says, "You have puffy cheeks, blood-shot eyes, slurred speech; you look dissipated. Any reason you can give?"

Charley says that he feels fine.

Shock describes the appearance and symptoms of the classic alcoholic and asks if Charley sees any parallel. Charley replies he sure as hell doesn't. He is not an alcoholic. He takes a few drinks now and then, but he can hold his liquor with the best of them.

Shock replies, "OK, Charley, I hope that you're all right, but your appearance concerns us; it's limiting your usefulness, we want improvement, and we will be watching your performance."

Charley backs out of Shock's office. "OK," he says, "but I'm no alcoholic."

As Shock and Chen agree later, alcoholics must recognize their problem, and until that happens, not much can be done for them. As Chen approached a decision to fire Charley for lack of performance, Charley resigned for another position, prompted by Chen's distrust and follow-on supervision.

Shock reviews the case with Watts and asks what his thoughts are.

Watts admits that alcoholics present a difficult problem and that such a person needs help. He says that Shock should have determined if the company has an employee assistance program. If so,

Charley could have been assigned to a rehabilitation program. Of course, if he rejected help, then his performance would determine his fate. Unfortunately, if an alcoholic is terminated or leaves, the problem gets passed to the next employer.

Watts also says that this is one of those cases where potential legal problems get tangled with personnel problems. Watts says Shock should document everything and should turn to the specialists in the personnel department for guidance if it happens again.

Shock says, "It's after five. How about a short one?"

Questions

- Would you have handled Charley differently?
- Why not check with personnel before the meeting with Charley or have someone from that department sit in?
- What if the employee had been suspected of using drugs?
- What if Charley had not resigned?

Good Today but Gone Tomorrow

Shock is looking at a résumé one day when Watts drops in to see how things are going. Shock says that Harry, who is now head of the electronics area, is in need of someone with an electronics background and experience as a computer programmer. The person being considered has the appropriate background and experience. He hands Watts the résumé and asks for his opinion.

The candidate has a B.S. degree in electrical engineering with reasonable training in electronics. He spent a year and a half as an electronics engineer with a local company when he decided to attend an industrial training school to take computer programming. He then spent one to two years each at six or seven different companies. The reasons for the job hopping included the company is going belly up, an offer of a better position, work that wasn't challenging, and so on.

Watts has a definite opinion: "If you hire this person, you'll have him for one to two years and then he'll get restless and move on. I don't think an employee pays out in that amount of time, and unless he says he will stay for at least five years, I wouldn't hire him. His repetitive moves suggest a pattern. His reasons for leaving sound like many I've heard before, and if you are really going to consider him, I suggest checking with his former supervisors. I see that he doesn't mention the name of his last supervisor."

Shock remains unconvinced. The applicant is still young. He is obviously energetic and ambitious. Maybe he just hasn't found the right spot. Shock is tempted to call the person in for an interview until Watts says, "You can't change the spots on a leopard."

Questions

- How do you distinguish between valid reasons for leaving a job and those contrived by a drifter?
- Do references given in the résumé include the candidate's superior for each position over the last five years?
- Are the references good friends or are they supervisors?
- Do times of employment and salaries given in the résumé match with those determined in reference checks?
- Would you verify all degrees, whether relevant to the job or not?

Publications versus Creativity

Shock asks Watts to take a look at another candidate. He's a Ph.D. electrochemist that Brown thinks would be a good addition to his fundamental alloy investigations staff. Shock hands Watts a résumé that would choke a horse. It extols the candidate's virtues in great detail. He received his doctorate from a reputable university and then stayed on for three years of postdoctorate work. He took a position with an oil company and did some very fundamental research. From there, he went to a contract research institute and for five years did research on government contracts until the money ran out. He has over 100 papers with titles confusing to the layman chemist. No patents.

Watts flips through the pages a second time before he comments. "He seems highly trained and experienced, but I doubt he would find satisfaction in this environment, since there are limited publication opportunities; this would eventually tick him off in the long run, especially if he thinks of publication as a survival tactic or as a reward. It's also doubtful that he would get the stimulation needed from colleagues, with all due respect to your staff, Shock. Your area of interest could be too limited for him, but he would make a good consultant. He probably should be with a research institute, a university, a government lab, or an industrial lab with a long-term commitment to electrochemistry. But most important, although he is an excellent specialist, he probably would not provide the creativity needed here. Note that he has no patents."

Shock wonders why the guy has an interest in his group, as he puts the résumé aside.

Questions

- Does the number of papers give a hint as to the person's creativity?

- Do patents give a hint as to a person's creativity?
- What is the basis for creativity in a multidisciplinary world?
- If the electrochemical projects are several and are seen as continuing over time, should Shock consider this candidate as a person on a technical ladder?
- Does the size of a group have a bearing on the feasibility of a technical ladder?

References

109. J. H. McQuaig, *How to Pick Men,* Frederich Fell, New York, 1963.
110. P. Balderston, P. Birnbaum, R. Goodman, and M. Stahl, *Modern Management Techniques in Engineering and R&D,* Van Nostrand Reinhold, New York, 1984.
111. M. Boyd, "How to Really Talk to Another Person," *Parade Mag.,* Feb. 19, 1989, pp. 14–15.
112. M. F. Wolff, "Motivating Research Scientists to Reach for the Eagles," *Res. Mgmt.,* September–October, 1986, pp. 8–10.
113. L. W. Porter and E. E. Lawler, III, "What Job Attitudes Tell about Motivation," *Harvard Bus. Rev.,* January–February 1968, pp. 118–126.
114. ———, *Managerial Attitudes and Performance,* Dorsey-Irwin, Homewood, Ill., 1968.
115. D. J. Cherrington, H. J. Reitz, and W. E. Scott, Jr., "Effects of Contingent and Noncontingent Reward on the Relationship between Satisfaction and Task Performance," *J. Appl. Psych.* **55**(6):531–536 (1971).
116. A. H. Maslow, *Motivation and Personality,* Harper & Row, New York, 1954.
117. M. Maltz, *Psycho-Cybernetics,* Prentice-Hall, Englewood Cliffs, N.J., 1968.
118. P. F. Drucker, *Management Tasks, Responsibilities, Practices,* Harper & Row, New York, 1974.
119. C. N. Greene, "Causal Connections Among Managers' Merit Pay, Job Satisfaction, and Performance," *J. Appl. Psych.* **58**(1):95–100 (1973).
120. M. K. Badawy, "Industrial Scientists and Engineers: Motivational Style Differences," *Calif. Mgmt. Rev.,* Fall 1971, pp. 11–16.
121. G. E. Pake, "Motivation, Nurturing, and Rewarding R&D People," *Proc. of the Conference Board National Conference on Research and Development: Key Issues for Management in 1986,* New York, February 1986.
122. C. Margerison, "Managing the R&D Group," *Mgmt. Decision* **16**(18):52–63 (1978).
123. G. E. Manners, Jr., J. A. Steger, and T. W. Zimmerer, "Motivating Your R&D Staff," *Res. Mgmt.,* September–October 1983, pp. 12–16.
124. W. E. Souder, "Award Programs for R&D Personnel," *Res. Mgmt.,* November–December 1985, pp. 13–18.
125. ———, "Award Programs for R&D Personnel," *Res. Mgmt.,* November–December, 1985, pp. 13–18.
126. L. R. Jauck, "Tailoring Incentives for Research," *Res. Mgmt.,* November 1976, pp. 23–27.
127. W. J. Turner, "Stimulating and Rewarding Invention—How the IBM Awards Program Works," *Res. Mgmt.,* July 1979, pp. 49–52.
128. R. Shapiro and S. Globerson, "An Incentive Plan for R&D Workers," *Res. Mgmt.,* September–October 1983, pp. 17–20.
129. J. C. Mower and D. Wileman, "Rewarding Technology Teamwork," *Res. Mgmt.,* September–October, 1989, pp. 24–29.

Suggested Readings

Badawy, M. K.: "Managing Human Resources," *Res. Techn. Mgmt.,* September–October 1988, pp. 21–35.

———: "One More Time: How to Evaluate Engineers," *IEEE Transactions on Engineering Mgmt.,* May 1978, pp. 37–42.

Carroll, P. J.: "The Link between Performance and Strategy," *J. Bus. Strategy,* Spring 1982, pp. 3–20.

Clark, C. H.: *Idea Management: How to Motivate Creativity and Innovation,* AMACOM, New York, 1980.

Coffin, R. C.: "Tailoring Incentive Bonus Plans for Special Situations," *Compensation Rev.* **7**(4):16–30 (1975).

Drews, T. R.: "Motivational Factors Relevant to the Performance of R&D Employees," *Ind. Mgmt.* **19**(4):5–11 (1977).

"How Companies Can Lengthen Their Sights," *Business Week,* June 30, 1980, pp. 92–94.

Keller, R. T., and W. E. Holland: "Toward a Selection Battery for Research and Development Professional Employees," *IEEE Transactions on Eng. Mgmt.,* **EM-26**(4):90–93 (1979).

McConkey, D. D.: "The Jackass Effect in Management Compensation," *Bus. Horizons,* June 1974, pp. 81–91.

Miller, E. C.: "Top- and Middle-Management Compensation—Part 2: Incentive Bonus and Merit Increase Plans," *Compensation Rev.* **8**(4):33–46 (1976).

Naor, J.: "How to Motivate Corporate Executives to Implement Long-Range Plans," *MSU Business Topics,* Summer 1976, pp. 41–49.

Parke, E. L., and C. Tausky: "The Mythology of Job Enrichment: Self-Actualization Revisited," *Personnel J.* **52**(5):12–21 (1975).

Rantfl, R. M.: "Training Managers for High Productivity: The Hughes Approach," *Natl. Prod. Rev.,* Spring 1984, pp. 170–180.

Rotondi, T., Jr.: "Effectiveness and Motivation among Professional Personnel—Part 2, Managerial Implications," *Ind. Mgmt.,* **V**(2):1–4 (1975) and M. Sashkin, *A Manager's Guide to Participative Management,* AMACOM, New York, 1982.

Scanlon, B. K.: "Creating a Climate for Achievement," *Bus. Horizons,* March–April 1981, pp. 5–9.

Sherrid, P.: "New Targeting for Executive Pay," *Fortune,* May 4, 1981, pp. 176–184.

Chapter

6

Control Activities

Questions answered in Chapter 6

- Where do managers obtain control measures for monitoring and reviewing their staff's activities?
- Why must managers review the planning process?
- What is considered during project reviews?
- What are reasonable techniques with which to control and monitor project time and costs?
- What is included in the budget review?
- How and why is documentation important to innovation?
- What is a good procedure for personnel reviews?

The strategic technical and operating plans and the follow-on operating plans provide the basis for managers' control of progress. Included in the strategic plan are a mission statement, goals, and strategies to reach goals. This blueprint determines the type of research that can be undertaken and sets the direction for the group's efforts. Managers can evaluate projects proposed during the year against this blueprint. Included in the operating plan are a portfolio of projects, each of which has goals, overhead and capital costs, and discussion subjects such as technical and market trends. Some of these, such as project costs, must be monitored (overseen continually), and others, such as technical results and profitability, will be reviewed (by periodic or ad hoc assessment) at a milestone. Also, a project portfolio is dynamic, and during the year managers will cancel some projects, transfer others, and add new ones. Each such action requires review and approval by managers. In addition to project costs, overhead items are monitored in order to meet the budget. Personnel must be evaluated periodically for progress on goals versus results and on self-improvement activities. These and other activities over which managers must have control are discussed in the following text.

The Planning Process Review

In Chap. 3, I described the planning process as flexible and adaptable to your needs. Because of this flexibility, the planning team should review the process, especially the year following its first use, to determine how it might be improved to fit their current circumstances. Some questions to ask are:

- Can the process be streamlined; that is, are there changes in the process that could be made to produce more results with less effort?
- How can data be gathered more efficiently?
- Have there been omissions in the planning?
- What can be deleted or deemphasized?
- Should there be more coordination with other groups?
- Is the report to management adequate? Can its preparation and content be streamlined?

Review the process as well as results versus the plan.

Project Progress Reviews

In the annual operating plan, all projects are posted with target times and costs for various milestones, results from various project evaluations, market potentials, profitability analyses, market share estimations, and so on. The manager must monitor such forecasts and review them periodically, probably once a week, with a more formal and intensive review at the end of each milestone. Of course, a good professional or manager will ask for review should some new problem or success suggest change in direction, rate of effort, or content of the work. Chapters 7, 8, and 10 describe some useful review techniques. Other relevant information is given in Chap. 4.

Project reviews are a valuable part of R&D.

In weekly project review meetings with R&D managers, as head of R&D I like to know, first, whether we are dealing successfully with constraints—technical, market, and legal. Second, as I listen to my staff, I am on the alert for any personnel problems. Third, the subject of lack of equipment or equipment failure usually arises, and I must consider refurbishment versus new equipment (lease or purchase) versus use of outside service equipment. Fourth, I want to see any new numbers from decision analysis on the benefit-to-cost ratio, internal rates of return for our own and the competitor's product, ROI, market share analysis, and other profitability analyses. Fifth, I want to know what communications we have had with downstream groups and what their reactions were. I then want to go into what will be done in the next period and what results can be expected.

Have your antennae out for unspoken problems.

We will also review time and costs. In addition, there are miscellaneous yet important items that may be brought up, such as safety issues, the need for an outside market research study, news about an emerging technology or a new piece of equipment being offered, some gossip on a competitor, the chance to get a good seminar speaker, and so on. We may also want to find out why the johns aren't clean, why Joe can't get to work on time, where we are to get funds for the Christmas tree, and so forth. The whole meeting should not take more than 1 hour if everyone comes prepared; no meeting should be allowed to go beyond 2 hours.

Decisions will be ongoing during these discussions so that no one is in limbo. We want all managers to have the cross-fertilization provided by ideas from everyone present.

Consider data and alternatives and make timely decisions.

As a project progresses, constraints and how they are being answered receive top priority in scheduled reviews. The R&D manager sees that the project leader seeks expert help to overcome any technical problems. The investigator reviews, interprets, and reports project results at each milestone. No project should escape this ritual of data review, interpretation, and reporting. The project team will learn by it, colleagues can best assist by looking at all the data, and management can interpret progress versus funds expended. Many times the investigator will find voids in R&D while digesting and interpreting raw experimental data.

It is often the case that periodic reviews are not enough. There are times when R&D hits knotty situations or investigators cannot explain results. In such cases, managers should require that the investigator write a report that sums up and interprets the work. The investigator then gives the report to colleagues for critique and constructive suggestions. Neglecting such reviews often leads to the accumulation of useless data, wheel spinning, and the disenchantment of management. For example, too often an R&D group's enthusiasm entices it into developing a product when there is *insufficient understanding of the fundamental cause and effect relationships* involved. If an adequate understanding is not achieved, you are liable to file for patents, later discover a relationship that has a broader and more generic meaning, and then desire to refile perhaps in the face of an issued patent that now becomes obstructing prior art. Managers must ask such questions as what controls the function of the product? how do physical phenomena affect changes in function? and so forth. If such voids had become evident in a project review, the research effort of the project would have been directed toward obtaining such understanding. Without review, wheels can spin merrily, and trouble will surely ensue. The responsibility for such reviews lies with the project leader and, of course, the area manager.

Financial Reviews

Time and cost monitoring and review

On large projects, managers may find it useful to use the critical path method (CPM) or program evaluation and review technique (PERT) to evaluate which series of tasks is critical, to estimate the time involved to complete the project, and to facilitate communication to management. The simple project shown in Fig. 6.1 reveals that the route involving exploration of tougher substrates and component A is the longer and is therefore the critical path. It also shows that reducing the length of time to choose a substrate for component B would be beneficial and should be given special attention. Time to complete the project is shown in calculations that follow.

To estimate task times such as in the network of Fig. 6.1, obtain estimates of the earliest (L1), likeliest (L2), and longest (L3) times to completion from the project leader or investigators. Assuming a bell-shaped distribution of estimated times, you can then estimate the expected times within which each task would be completed with a 50 percent chance and the time of the total project by:[22,130]

$$ET = [L1 + (4 \times L2) + L3] \div 6 = \text{expected time}$$

$$D = (L3 - L1) \div 6 = \text{standard deviation of a task}$$

Task	L1	L2	L3	ET	D^2
Substrate selection	4	8	16	8.67	4.00
Coat components	1	4	6	3.83	0.69
Examine components	1	2	3	2.00	0.11
System test	2	3	5	3.50	0.25
Review	1	1	1	1.00	0.00
Total	9	18	31	19.00	5.05

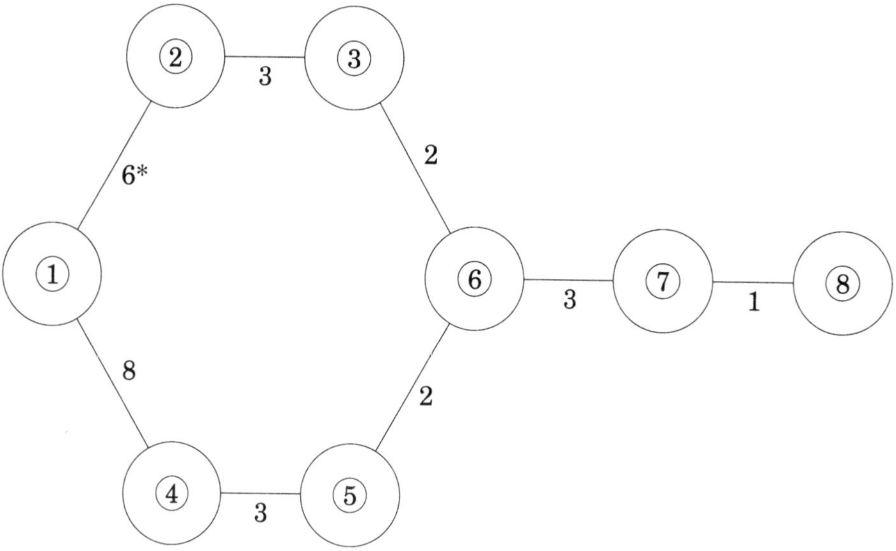

Fig. 6.1 Determine the critical path. 1: Start, 8: Conclude. 1–2: Explore coatings, 2–3: Coat and test. 1–4: Explore substrates, 4–5: Coat A and test. 3–6 and 5–6: Examine, 6–7: Test system. 7–8: Review. *Time, in weeks.

By such mathematics, the project would be completed in 19 weeks at a 50 percent probability. A sum of the likely times (L2) is 18 weeks. The task that might have the greatest affect on timing is brought out in the D^2 column. The standard deviation for the project time would be $5.05^{1/2}$, or 2.25.

The project leader must track project costs. Although a central project cost accountant can do this in close coordination with the project leader, it is best if the project leader in R&D takes this responsibility. It is also to the project leader's advantage; if he or she is called on for a report, the current balance will be at hand. Although a statement of project budget minus expenditures gives a current balance, a visual presentation is often the best technique to show management how the project is doing. For example, costs may be heavy at first or at some period during the project and light at other times. Curves of expenditures versus time for anticipated and actual costs often communicate the circumstances better than a running balance (see Fig. 6.2). Sometimes the best technique is to use both statement and visual aids.

Profitability review

Managers are concerned with profitability of projects, purchase of equipment, license or purchase of outside inventions, proposed acqui-

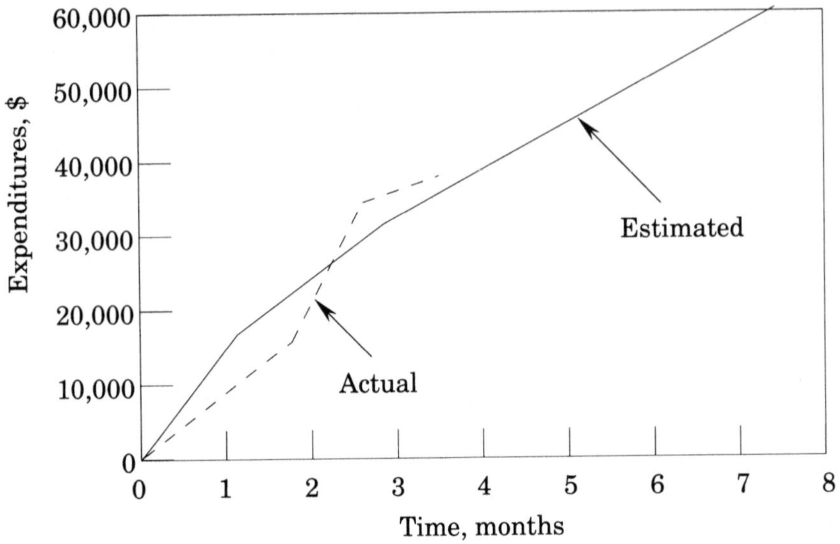

Fig. 6.2 Costs versus time.

sitions, licensing or sale of patents, and so forth. Each action is reviewed and approved by the vice president of R&D or corporate offices, depending on the subject. Profitability must be uppermost in a manager's mind when expending, or proposing to expend, company funds. It is of such importance and complexity that I have deferred discussion until Chap. 10. In Chap. 10, I also discuss return on investment, internal rate of return, payback period, and other measures of profitability. Calculation and use of the measures are demonstrated.

Profitability is the ultimate goal.

Budget review

R&D groups usually prepare two budgets annually, the operating budget and the capital budget. The operating budget is a compilation of project expenses, consultants' expenses, travel, outside services, and all personnel costs for each research area; administrative costs; service costs; taxes; depreciation; and all other costs of keeping people, equipment, and facilities. The capital budget is formed by negotiation from wish lists received from the research areas (see also Chap. 3).

Budgets must be justified.

During the preparation of the annual report, each research manager assesses all projects and examines finances relative to accomplishments. He or she looks at what has been spent and what must be spent to reach certain milestones and reconciles any differences between these costs and original estimates. If investigators have given these costs continual attention and periodic review during the year, the manager's annual review and report should proceed smoothly.

Documentation of New Concepts and Data

Good record keeping is basic to successful innovation, and it is the duty of all managers to see that their staffs recognize the need for good records. Managers must continually monitor records to determine that they are being well kept, properly witnessed, and current. Such monitoring is an essential control activity—lost patent rights being only one result of inadequate records.

Make original records timely and complete.

Investigators record all ideas for possible patent actions and for the idea evaluation exercise described in Chap. 8. If investigators don't record an idea in a notebook at the time of its conception, something of greater interest will come along and the idea may be lost forever. Or they may themselves shoot down the idea without giving discussions with colleagues a chance to evolve a synergistic result.

Investigators should enter new ideas into the project notebook or, if this is not appropriate, into a general idea notebook. I have also promoted the use of a one-page form called the *idea record form* (see Exhibit 6.1), which provides space for a description of the idea and its prior art. Signed and witnessed, this form becomes a useful legal document as well as one that can be used in idea evaluations. Use of disclosure forms to record an idea initially is wrong; such forms are too long, and many creative people are loath to make the required effort. Use a disclosure form after the feasibility study.

All professionals and managers should have a bound notebook for each project, and managers should instruct their staffs on how to enter data and information into notebooks properly. Date, number, and title each page (see Exhibit 6.2). Leave no blank space on a page; cross through unused space. Use an indelible pencil or pen (for possible legal purposes). Put original data into the notebook, not on scratch paper. Take explicit notes and describe events and aims; include your observations and thoughts. Note any peculiarities; small, seemingly unimportant observations may be important later.

Record all observations no matter how trivial; they may be vital later.

IDEA RECORD

(Use other side if necessary—cross out all unused space)

NOTEBOOK REFERENCE ______________________ COPY NO. ________

GENERAL SUBJECT ______________________

DESCRIPTION:

SIGNIFICANCE AND PRIOR ART:

Discussed with me on ______________________ ____________

Date Signature Date Signed

Inventor's signature ______________________ ____________

Read & understood by me ______________________ ____________

Witness

______________________ ____________

Witness

Exhibit 6.1 Idea record.

Every originator of a notebook entry must sign and date each page and have every page signed and dated by a witness. Drawings can be made on a page or attached to a page with tape. Staple separate worksheets and photographs to a page. If routine work requires forms for original data, these must be signed, witnessed, and dated. Include cross-references to separate worksheets, chart recordings, graphs from recorders, and the like in the notebook. Bind data forms and graphs for safekeeping.

Notebooks are the property of the company, and when a project is

DATE TITLE PAGE NO.

Description of the experiment ____________________

Data ____________________

Interpretation of data ____________________

A concept based on the results ____________________

[Cross through unused space.]

Witness:
Signature ____________________
Date ____________________

Investigator:
Signature ____________________
Date ____________________

Exhibit 6.2 Format for a notebook page.

completed, they should be returned to central files. Also, on termination all employees must certify that they have turned in all their notebooks and original records before they receive final benefits.

This task of record keeping is the most often violated procedure among the technical staff. I have no surefire method for ensuring that everyone provides well-kept records. A manager's vigilance will have the most influence in this regard. Remind your staff that a notebook's value often lies in its future use as a possible legal, as well as technical, document. Warn them that when they are on a witness stand and the company loses because of their shoddy notebooks, they will be mightily embarrassed. If you have a particularly recalcitrant employee, add the notebook requirement to his or her annual review considerations.

Personnel Review

Personnel appraisal is an important control function. Staffs and managers of each area must be evaluated on their performance, personal growth, how they dealt with problems, and how they met personal and company goals. What they hope to accomplish in the future is also discussed. The result of each interview is agreement between manager and subordinate on the evaluation of the subordinate based on these criteria and on goals for the future. This means that what was done in the past year must be compared with what was agreed upon at the last appraisal and a new agreement made on future goals.

The employee and manager discuss results versus goals.

In the appraisal meeting, you are interested mainly in two things: results accomplished versus previously agreed upon goals and personal improvement. Use these guidelines:

- Put the subordinate at ease. This must be a cordial two-way exchange of information on an important asset—him or her.
- Ask questions rather than make appraisal statements unless they summarize what the subordinate has said or unless the subordinate evades the issues. Let the subordinate talk freely. Be relaxed and take the time needed to get a complete appraisal, with both sides satisfied that everything that could be done was done.
- Start with results versus goals. Ask about successes and failures. Ask why both happened and see if any patterns of behavior are ap-

parent. Discuss whether alternative actions might have produced different results.

- Ask what has been done regarding self-improvement. A subordinate's growth is important to both him or her and the company.
- Ask what more can be done by the group and the company to promote technical success and his or her personal growth. Discuss his or her suggestions to ensure mutual understanding. You may uncover some thoughts and reservations about the group or company that are not warranted, and these can be clarified.
- Ask about goals for the next period. Mutually decide on what is to be accomplished. It is important that the subordinate set the times for accomplishment.
- After the meeting, record the results and fill out the personnel forms.

The company requires that managers fill out standard forms and return them to the personnel department. Often, these are too long, tedious, and superficial, but they must be a part of the appraisal process, since they form equivalent appraisals among like pay grades.

Although such meetings are usually held annually, during the year some subordinates may want to discuss how you think they are doing, and your door should always be open to them. For those persons with whom *you* want to discuss their progress or, more likely, lack of progress, meetings should be held at once. See also the section "Obtain and Retain Only the Best Staff" in Chap. 5.

The Annual Operating Report

All reviews come together annually in the preparation of the operating plan. Project reviews provide information on technical progress; there is a discussion of constraints and a reconciliation of actual versus proposed time and costs. Any proposed changes in the strategic plan are identified. Technical and market trends are described. The portfolio of projects for the coming year is presented with time and costs. A financial review is given for the past and coming year, including project and overhead costs. A capital budget is proposed.

Management sees in this report the results of the group's planning, implementing, leading, and controlling activities during the year.

Summary

- Control of progress toward successful innovation starts with the strategic technical and operating plans.
- The strategic technical planning process warrants periodic review to improve efficiency.
- Project reviews, held periodically at milestones and on an ad hoc basis, are to:

  Discuss constraints.
  Decide on personnel problems or needs.
  Describe equipment needs.
  Evaluate recent analyses on competitive status.
  Evaluate communications with other groups.
  Describe and discuss future work.
  Monitor time and costs.
  Review miscellaneous items such as safety.
  Formalize decisions.

- Project time and costs must be monitored continually to avoid or cope with overruns.
- The annual budget submitted with the operating plan includes project and overhead expenses and capital needs.
- During personnel appraisals:

  Put the subordinate at ease.
  Use questions to obtain an appraisal of performance.
  Identify accomplishments versus goals.
  Determine what has been done on self-improvement.
  Obtain suggestions on how the group could be improved.
  Help the employee set goals.
  Record results of the meeting.

- Idea and data documentation describe technical progress and proprietary aspects and are essential to legal proof of ownership.

Cases and Incidents

Don't Step to the Plate Unless You're Ready to Swing

Shock's group is involved in some water pollution hardware that could have significant impact on pollution aside from providing new profits to the company. The hardware development has progressed reasonably well, but because of its high visibility in the company, many evaluation steps were shortened or eliminated in the push to get a working prototype. Passage of federal standards is a necessary constraint and emphasis is put on passing those rather than on obtaining an improved understanding of how the hardware operates fundamentally. The CEO tells Shock they should try to sell or license the hardware, so Shock contacts selected companies.

From the several contacted, one company is picked, but negotiations go nowhere. The company vacillates, always wanting more information. Finally, corporate decides that the R&D group should enter into a long-term test with a federal agency as a move toward getting industrial acceptance. The CEO asks Watts to arrange such a test. Watts and Shock visit a friend in a local agency in order to get a feel for the steps they should take. Shock reviews the technical status of the hardware.

Watts's friend says, "It sounds as if you have a potential solution to an important problem but not the commercial hardware. You have to realize that once these tests are started, you can't stop one of them to do some research; the test grinds inexorably to a conclusion, and the result, good or bad, will be public information. So, my advice is don't step up to the plate unless you're ready to swing."

At this point, he tells them the steps to go through and the people to see.

Watts tells Shock, "This is good advice to any R&D group. Don't talk about an idea prematurely, don't transfer technology before constraints are studied, and to paraphrase an advertisement, don't sell anything before its time."

Shock adds, "Or you've struck out."

Questions

- Is it good policy to subvert fundamental study of operations to obtain operating results?
- Can a product be sold without a clear understanding of its operational fundamentals? What about patent validity?
- Should prototype hardware be submitted to public test? Under what circumstances?

Lack of Safety Measures Can Ruin Your Whole Day

Shock and Watts are talking one day when the subject of safety comes up. Watts says, "Shock, concern yourself with safety practices in this group before too long. I am not aware of any unsafe conditions or practices, but safety is a fetish of mine, and I am constantly concerned with such matters."

He continues, "There is the story of the young man running an experiment with a hazardous gas that he had not discussed with his supervisor or colleagues. The particular chemistry is notorious for unforeseen explosions. He didn't consider adequate safety precautions; he had an idea and wanted to see if it would work! Fortunately, he was stopped before the experiment went too far.

"There is also the story about the person who, as was routine, hooked up a gas bottle to equipment thinking it was an inert gas; however, someone had switched bottles. After the explosion, it was recognized that the equipment had been hooked up to a highly combustible gas. And there was the glass reactor which blew but fortunately was in a concrete bunker. Glass splinters found their way around the bunker walls and, turning 90 degrees, embedded themselves in a concrete block wall 20 feet away.

"Then there was the case in which an ozonide synthesis got away from an investigator, and a savior came to the rescue with a carbon dioxide bottle. In another instance, a preparation of sodium sand also got away, with sodium slurry coming out the top of the condenser, and the heating mantle could not be removed from the reactor.

"A high-pressure autoclave experiment went bad but was contained inside the bunker. Unfortunately, the investigator had mounted his instruments on a plywood panel affixed to the bunker wall. Splinters flew from the panel, injuring him.

"You must see that managers monitor and that professionals are continually vigilant to safe experimental practices. A safety manual and safety equipment should be available, if not issued, to each staff member. As head of R&D, you should appoint a safety inspector, and if the appointment is not permanent, staff members can rotate. The technical staff should be completely familiar with all safety equipment. For example, let the local (or company) fire department show your staff how to use fire extinguishers. Show movies on fire safety. Include safety as one subject for your seminars. Have your corporate safety inspector talk to your staff members."

Shock grins as he says, "Those are great suggestions. I'll fire up the staff."

Questions

- Who is your safety inspector?
- Do you have a copy of the company's safety booklet?
- Do you concern yourself with safe practices? If so, does your neighbor?
- If you were a safety inspector for your group, what conditions and practices would you target for change?

References

130. J. M. Nevison, *Executive Computing,* Addison-Wesley, Menlo Park, Calif., 1982. See also texts on PERT or CPM, such as R. D. Archibald and R. L. Villoria, *Network-Based Management Systems (PERT/CPM),* Wiley, New York, 1967, and R. W. Miller, *Schedule, Cost, and Profit Control with PERT,* McGraw-Hill, New York, 1963.

Suggested Readings

Meinhart, W. A., and J. A. Pederson: "Measuring the Performance of R&D Professionals," *Res. Techn. Mgmt.,* July–August, 1989, pp. 19–21.

Part

3

Methods and Techniques

In Part 3, I discuss a few of the methods and techniques useful to technical managers during their activities. This is by no means an exhaustive coverage but is a brief presentation of those considered to be most useful and those that complement the previous chapters.

Chapter

7

Techniques to Facilitate Planning Activities

Questions answered in Chapter 7

- What are some useful technical forecasting methods?
- What is a technical planning study?
- How do you formulate a scenario?
- How is decision analysis applied to evaluations of potential projects and project portfolios?
- How can matrix analysis be used in planning?

In this chapter, I discuss the techniques that I have found most useful during planning of R&D. You don't have to be an expert at technical forecasting or decision analysis, although you can be, but you want to have enough knowledge of planning techniques to make your activities easier and your results more complete. The "Bibliography on Technological Substitution" at the end of this chapter and the reference section provide additional detail for these and other techniques.

An arbitrary selection of techniques is presented.

Technical forecasting is used in strategic planning studies, in technical planning studies, or in specific cases of interest. It is important to select combinations of forecasting methods that enhance the accuracy, and hence utility, of the result, for example, trend extrapolation in combination with expert opinion, Fisher-Pry along with Blackman, and so forth.

Decision analysis is worth your extensive study, especially for its use in portfolio analysis. Many R&D managers use matrix analysis in their planning process; you should determine its usefulness to your group.

Technical Forecasting

There are some 200 technical forecasting methods. The simpler ones discussed here are those I have found most useful, especially to the part-time forecaster or R&D manager. Descriptions of the methods will be very brief, and the reader will wish to expand on those of interest by reading the references and bibliography and by attending courses such as those presented by Technology Futures. Texts have the advantage of depth and extensiveness of treatment; courses have the advantage of offering workshops and information about the latest accomplishments in forecasting. Forecasts are an input to the planning process and allow you to identify the possible impact of change on corporate business and the directions your R&D should take.

The technical progress or performance of industrial products typically follows an S-shaped curve of performance versus time (see Fig. 7.1), whether the measurement be of a product life cycle, a technology life cycle, or some technical or economic performance parameter. Plotted on arithmetic scales, S-shaped curves reflect a slow start followed by an exponential growth and then a leveling off as the curve asymptotically approaches some natural limit. The exponential period can

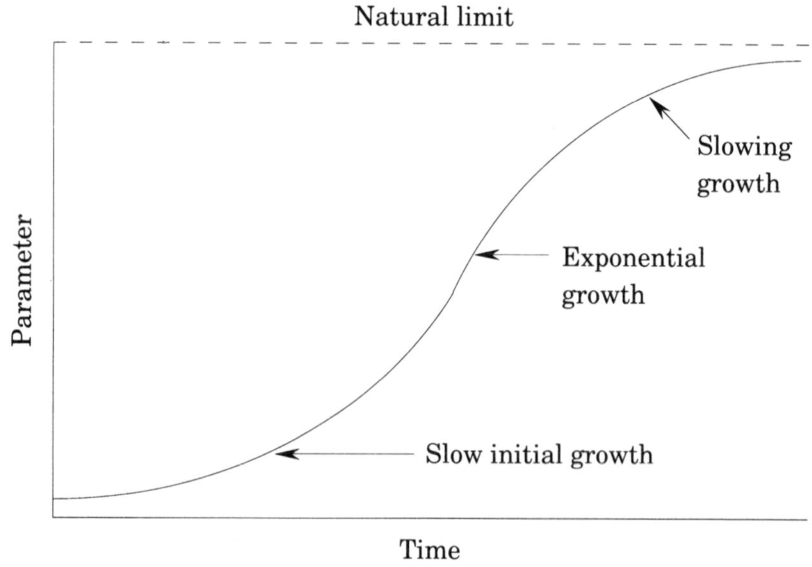

Fig. 7.1 The S curve.

have various slopes. A new technology would exhibit another S-shaped curve, presumably showing a higher product performance trending beyond the natural limit of the first curve and asymptotically approaching a new and higher natural limit.

A product life cycle begins with considerations of market entry; on introduction of the product into the market, the cycle goes into an exponential growth period; growth levels off and may fluctuate or decline slightly over a period of years; and then the growth goes into an exponential decline as a new product takes over.

Estimate where your products are in their life cycles.

Your choice of forecasting techniques depends, in part, on the amount of data available for a specific product.

Forecasting with few or no data

Delphi procedure. The Delphi is a forecasting technique developed by a Rand Corporation team[131] in the 1950s and 1960s and uses expert opinion. A key feature is the anonymity of the experts taking part in the exercise. In essence, you ask each expert for a prediction. You tabulate the predictions and return all the results to each of the participants. Each participant reviews the results and reconsiders his or her prediction. If the reconsidered prediction is outside the middle 50 percent, the participant provides arguments supporting the nonconformance. Each participant considers the nonconforming positions and arguments and then revises his or her prediction and argument. You make three more rounds and then collate the results. Use expert opinion to modify results from trend extrapolation and to aid in formulating input to scenarios.

Expert opinion is helpful to moderate trend extrapolation.

Substitution technology. This is a technique used when a new technology is replacing an existing technology. The Fisher-Pry method[132] provides insight into the substitution process and its rate; Fig. 7.2 shows a Fisher-Pry plot of the fraction of the new technology divided by the fraction of the old technology versus time. The method suggests that when the new technology has taken 20 percent or so of the market, you can reasonably forecast the future, and when the new technology has taken about 50 percent, it will go on inexorably to consume the total market.

With few data, the Fisher-Pry plot offers a useful forecasting tool.

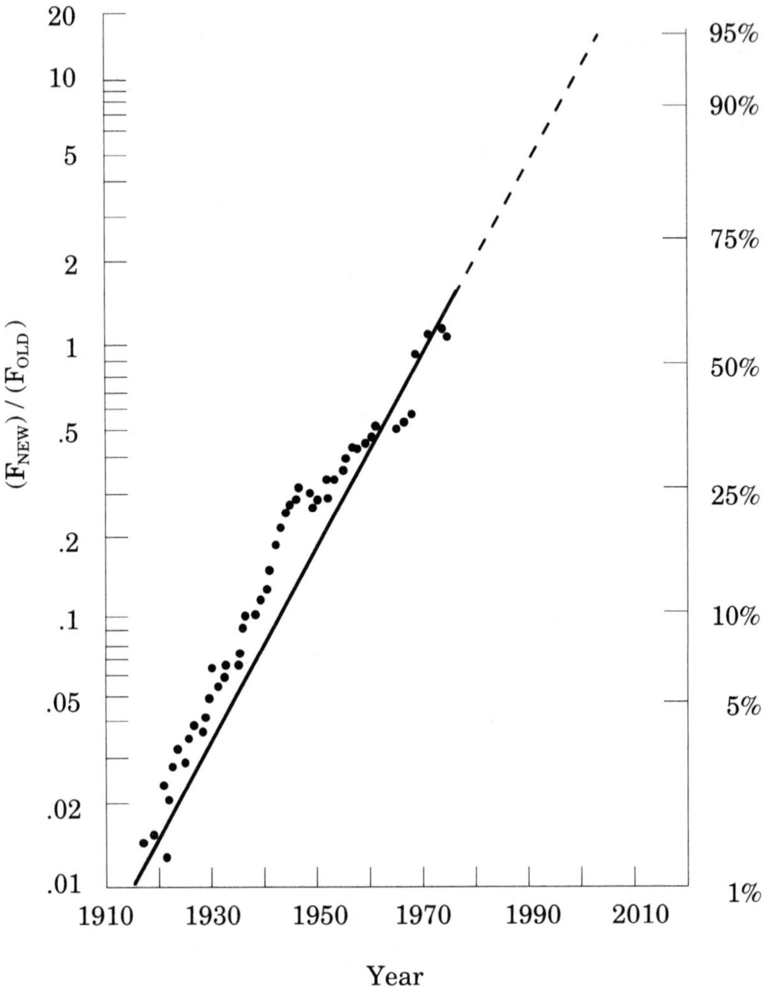

Fig. 7.2 A Fisher-Pry plot showing substitution of strip mining for underground mining of coal. *(Published by permission of Mr. Ralph Lenz.)*

Although there are many examples to validate these statements, environmental influences can change the conclusions. With the coal example (Fig. 7.2), you can propose several scenarios. One scenario is that strip mining continually substitutes for underground mining. Another is that a sudden demand for coal causes underground mining to expand quickly because the mines are there; in this example, you would see a temporary divergence until once again strip mining substituted for underground mining, and you are back on a curve having the original slope or even back on the original curve. In a third scenario, strip mining sites become more difficult to mine, transportation becomes more expensive, and, in essence, the economic comparison changes and the curve no longer becomes a valid depiction. There may be other changes that militate for continuing or ending a relationship.

The Fisher-Pry depiction makes us think through the various changes that might occur and makes us reach conclusions on how the future might look so that reasonable decisions can be made today. Although this is a valuable outcome of the Fisher-Pry technique, I strongly recommend that you compare the results of any such forecast with a forecast from a second approach. Technology Futures is offering a menu-driven software package that can be varied to fit the amount of data you have and provide forecasts in either graphical or tabular form.

The Fisher-Pry plot should make us think.

With as little as 2 percent market share, you can extrapolate historical data by the Fisher-Pry technique in order to "guesstimate" future inroads of a new product. With no historical data, you may still wish to estimate how a new technology may substitute for an old technology. Blackman[133] has seven equations (see Exhibit 7.1) which you can use to estimate a new product's future market share at time t, by relating benefits, expenses, and depreciation of old and new products. Blackman and an arbitrary selection of others who have provided valuable input to this area of substitution technology are listed in the special bibliography at the end of this chapter.

In the equations, Y is the purchase price, B is the estimate of annual benefits to the user, T is the tax rate, E is the estimate of expenses, D is depreciation, and j is the life of the product. Subscripts relate the old and new products. Calculate the benefits of the old product from Eq. 1. Equate the benefits for the new product to those from the old product and use this figure in Eq. 2. Calculate the rate of return of the new product. After determining phi (ϕ) from Blackman's Eqs. 3 through 6, obtain the market share from Eq. 7. For more easy manipulation, you can devise a computer program.

Problems with the method include difficulty in estimating the market limit L, differences in old and new product benefits, and the uncertainty of success. Attempts to overcome these led me to speculate on a second market substitution program, based on modifications of Blackman's equations. In this approach, calculate internal rates of return (IRR) of the old and new products, and compare the return for the new product with that of the old product; this gives you a profitability index. If this index is less than 1, the product is not competitive. The upper limit for the new product's market share is estimated by adaptation of a method by Souder and Quaddus.[134] Obtain expert opinion on the maximum perceived benefits of the innovation (VI) and the present benefits of the competitive product (VO). Divide the incremen-

$$1.\quad Y_O = \sum_{k=1}^{j} \frac{B_{o,k}(1 - T)}{(1 + ROR_{req})^k} - \sum_{k=1}^{j} \frac{(E_{o,k} + D_{o,k})(1 - T)}{(1 + ROR_{req})^k} + \sum_{k=1}^{j} \frac{D_{o,k}}{(1 + ROR_{req})^k}$$

$$2.\quad Y_N = \sum_{k=1}^{j} \frac{B_{o,k}(1 - T)}{(1 + \mathrm{ROR})^k} - \sum_{k=1}^{j} \frac{(E_{N,k} + D_{N,k})(1 - T)}{(1 + \mathrm{ROR})^k} + \sum_{k=1}^{j} \frac{D_{N,k}}{(1 + \mathrm{ROR})^k}$$

$$3.\quad S = \frac{\text{initial investment} \times 100}{\text{total assets of average purchaser}}$$

$$4.\quad \pi = \frac{\mathrm{ROR}}{\mathrm{ROR_{req}}}$$

$$5.\quad Z = 0.2221\,(I) - 0.3165$$

$$6.\quad \phi = Z + 0.530\pi - 0.027S$$

$$7.\quad In\left[\frac{m}{L - m}\right] = -\,In\left(\frac{l}{N_o} - 1\right) + \phi(t - t_1)$$

where m = market share captured at time t by the new innovation
L = upper limit of market share
N_o = market share captured when $t = t_1$
ϕ = constant governing rate of substitution
Z = an industry constant
π = a profitability index
S = an investment index
I = an innovation index
ROR = rate of return associated with the innovation
$\mathrm{ROR_{req}}$ = minimum rate of return required
Y_o = initial purchase price of old product
B_o = annual benefit to the user of the old product
T = tax rate
E_o = annual operating expense using the old product
D_o = depreciation expense of the old product
j = life of the product in years
Y_N, E_N, D_N = same as above but associated with the new product

Exhibit 7.1 Blackman's equations (*reprinted by permission of Elsevier-North Holland Publishing Company*).

tal increase in benefits by the benefits of the old product:

$$\frac{\text{VI} - \text{VO}}{\text{VO}} = F$$

Then obtain the limit of market share using the equation:

$$L = \{1 - [0.99 \times \exp(-F)]\} \times 100$$

Uncertainty is built into the program. If the internal rate of return is less than the cost of money plus 100 minus the probability of technical success for the new product, the venture is not profitable. For example, if the cost of money is 12 percent and the probability of success is 90 percent, the IRR would have to exceed 22 percent. This adaptation gave me results similar to those obtained from Blackman's program in a test case; otherwise, it is untried and untested.

Aside from its use in market research, a sensitivity analysis employing the Blackman method can be used to guide future R&D. For example, such analysis can aid you in deciding whether to concentrate on getting first costs or operating costs down and so on.

I suggest that the area manager, with help from the project leader, estimate market penetration over time for the specific market segment selected based on product and consumer characteristics and a history of previous product introductions. The above methods of estimating market share will be useful adjuncts to this procedure. Such market share estimates and sales should be reviewed with the marketing department and a consensus reached.

Forecasting with historical data

Trend extrapolation. In various forms, trend extrapolation is a simple and often used technique. However, a variable may consistently grow or decline over a long period of time, or it may show short-term seasonal or cyclic patterns. Seasonal variations may be in response to, for example, lower use during winter; a cyclic variation may be a response to the business cycle.

Trend extrapolation is a simple and useful technique.

In a time-series analysis, you plot the variable against time (for example, sales versus time) and use curve fitting to obtain the best straight line, exponential curve, natural log curve, or power curve fit. Hand calculator programs and software are available for this. Extrapolation yields the forecast.

If you are dealing with a product having a well-behaved growth pattern, trend extrapolation is satisfactory (for some arbitrarily limited number of years). A plot on semilog paper or the use of linear regression analysis (for which there are hand calculator and computer programs) will give data that can be extrapolated into the future to yield a forecast. However, you should obtain the protection of several approaches, and I advise use of some expert opinion, even that short of a full Delphi, with which to modify or shape the extrapolated trend line.

If use of a product depends on several variables, then a multiple linear regression analysis may be useful. These variables might be user's cash flow, costs of operating with the product, prices of the commodity produced with use of the product, gross national product, and so on. The typical equation in such an analysis,

$$Y = a_0 + a_1X_1 + a_2X_2 + a_3X_3 + \cdots + a_iX_i$$

assumes that the given relationship adequately describes the dependence of Y on the independent variables $X_1, X_2,\ldots,X_i$, with dependence expressed by the coefficients $a_1, a_2,\ldots,a_i$. Determine the validity of the relationship and the values of the coefficients from historical data. Then examine the equation to determine if you can simplify it by examining the standard deviation, error in the correlation coefficients, and the results of a sensitivity analysis. You then generate a forecast with the simplified equation.

Pressure curve analysis. Pressure curve analysis is also called *pressure curve forecasting* or the *rate-of-change method* and is a forecasting method used with products sensitive to business cycles.[135,136] You must ascertain the cycle pattern and the turning points to obtain two- to three-year trends and possibly to establish the beginning of a new trend line. Let's take the example of rigs used to drill wells for oil and gas (see Fig 7.3). For the period 1979–1982, trend extrapolation would give reasonable forecasts of active rig count (because the rate of change was fairly constant). That was a period when there was a strong upward swing of oil business activity. But during the sharp decline in activity during 1982–1983, trend extrapolation would give erroneous results. A method of monitoring the cyclic pattern for its strength and timing over a period of at least 10 years would have forecast this decline.

To damp seasonal variations, take monthly data and derive moving totals (or averages). Then ratio the current 12-month total to the 12-month total of the previous year. Plotted, this is your 12/12 curve. If you ratio the current 3-month moving total to that of the previous year, you have data for the 3/12 curve. Use the 3/12 curve to predict changes in the trend of the 12/12 curve; note that it generally leads the 12/12 curve by 4 to 6 months. The 12/12 curve tells you about changes in trend.

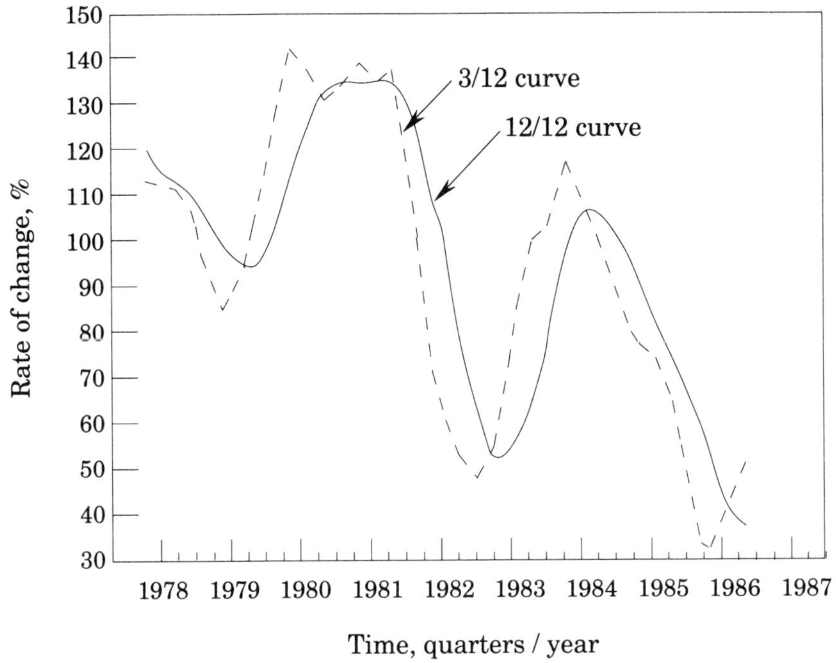

Fig. 7.3 Pressure curve analysis of active rig count.

You can now determine the cycle time between troughs and peaks from data in Fig. 7.3. With drill rigs, the time between peaks is about 42 months, and the rise in activity from 1979–1982 and the decline in 1983 were reasonably predicted from a 12/12 curve. However, peak-to-peak cycle times vary according to the industry technology and the length of the general economic cycles. (With rig count, troughs occur every 36–45 months.)

Get an initial fix on peak-to-peak times from historical data.

Cahners Economics[135] has a procedure to forecast the future by using 12/12 curves of such leading indicators as the Cahners' early warning indicator, The Federal Reserve Board's index of industrial production, and the U.S. Department of Commerce's index of 12 leading indicators. To obtain an approximation of monthly averages, lay the 12/12 curve of the leading indicator on the 12/12 curve of the product and match the peaks of the two curves, extrapolate the product curve, and back out the quarterly forecast data from the extrapolated curve.

Using the 12/12 values taken from the forecast portion of the curve, calculate the 12-month and 3-month moving totals:

$$12MT = \frac{12/12 \text{ value}}{100} \, 12MT \text{ (year ago)}$$

$$3MT = 12MT + 3MT \text{ (year ago)} - 12MT \text{ (quarter ago)}$$

where MT = moving total
12MT = twelve-month moving total
3MT = three-month moving total

Obtain monthly averages from the 3MT values. If they look reasonable, the forecast is done. If the values have some awkward swings, use seasonal and cyclic multipliers to "correct" them.

To obtain seasonal multipliers, take 3MT data for several years, sum the data by quarters, find the fractional value of each of these multiyear quarters relative to the total of all four multiyear quarters, and then to find the seasonal multiplier for a given quarter, divide its fractional value by that of the preceding quarter. Historical 3MT data are used in Table 7.1 to derive seasonal multipliers for active rig counts. It can be seen from the seasonal multipliers that there is relatively greater rig activity in the fourth quarter than in the preceding quarters.

A second way of forecasting is to extend the 3/12 curve based on what you anticipate the timing and magnitude will be for the 12/12 peak (or trough). The 12/12 curve is thus used as an adjustable template for finding the 3/12 curve. Multiply the 12/12 value assigned to the selected peak by some multiplier to account for the possible peak of the 3/12 curve. Note in Fig. 7.3 that the 3/12 peak is perhaps 5 to 10 percent higher than the 12/12 peak, so the multiplier is 1.05 to 1.10. Assign the resulting value to the quarter in which you anticipate that the 3/12 curve will peak before the peak of the 12/12 curve; in this instance, the 3/12 peak will be 3 to 6 months earlier than the 12/12 peak. You can now extend the 3/12 curve from where it might be to the new peak and then even to a new trough. If you take the fourth root of the 3/12 value, you will get an approximate average of the cyclic multipliers for that and the previous three quarters. Now you must adjust cyclic multiplier values until the 12/12 value desired is obtained in the quarter desired.

The 3/12 curve is an early indicator.

More explicitly, the previous quarter of 3MT multiplied by the seasonal and cyclic multipliers gives you the 3MT for the current quarter. If you subtract the year-ago 3MT and add the result to the quarter ago or previous 12MT, you get the current 12MT. This divided by the year ago 12MT gives you the 12/12 value. Do this for all values, and build

TABLE 7.1 Seasonal Multipliers for Active Rig Count

	Year and active rig count									
Quarter	1981	1982	1983	1984	1985	1986	1987	Total	Fraction	Seasonal multiplier
1	10483	12412	6818	7213	6504	4042	2411	49883	0.276	1.031
2	11471	9546	5751	6813	5615	2287	2340	43823	0.234	0.876
3	12371	7849	6447	7226	5783	2249	3023	44948	0.240	1.026
4	13308	7598	7751	7889	5686	2756	3436	48424	0.259	1.079
								187078		

a table that will show you when the resulting 12/12 values will peak and what the peak value will be. You might be a quarter or so away from where you want the future peak to occur, and the magnitude may not be exactly as assigned, so you must now iteratively adjust the the cyclic multipliers and recalculate the data until you have the desired value for the 12/12 peak and it is where you want it in time. Now you can obtain the average monthly values from your 3MT values. If everything looks reasonable, you may wish to round off the 3MT values and recalculate all data for the finished forecast.

A forecast for rig count by the second method is shown in Table 7.2. Note that all data in row 1 and data in column 4, rows 1–5, and column 7, rows 1–5, are historical; the rest are derived. The scenario used included the conclusion that the oil service business would not get out of its present doldrums until the late 1990s, or well past the period of this forecast. Although monthly values through 1988 look reasonable, those for late 1989 and all of 1990 appear lower than would be expected. The forecast must be monitored and redone periodically.

The rig count forecast shows the usefulness and the frailty of the method.

No forecast is guaranteed. It may be the best that you can derive at the moment, but you must update the forecast month by month. Should the 3/12 curve show a turning point at some point different from the one you picked, watch both curves carefully; if a rising 12/12 curve suddenly flattens out or drops while the 3/12 curve continues to decline, redo your forecast. The 3/12 curve, which need not be correct, acts as an early warning. Change may occur in the curves to the first forecasted point and you can expect change in the curves to the second forecasted turning point. Thus, any second forecasted turning point has an extremely low probability of being accurate.

The only way to gain facility in the use of pressure curve analysis is to practice it. If practice still leaves you inept, then contact Cahners Economics,* which has a detailed seminar on the subject and a manual to guide the uninitiated.

Levenbach and Cleary[137] determine cyclic patterns from deviations between the three-month moving average of seasonally adjusted data and the straight trend line through the data.

*Cahners Economics, Cahners Publishing Co., 221 Columbus Avenue, Boston, MA 02116.

TABLE 7.2 Forecast of Active Rig Count

Date	12MT	12/12	Year-ago 12MT	3MT	3/12	Year-ago 3MT	SM	CM	Monthly value
86:D	11334	48	23588	2756	48	5686	1.079	1.148	919
87:M	10219	48	21126	2927	72	4042	1.031	1.030	976
J	10599	60	17798	2667	117	2287	0.876	1.040	889
S	11250	79	14264	2900	129	2249	1.026	1.060	967
D	11874	105	11334	3380	130	2756	1.079	1.080	1127
88:M	12285	120	10219	3338	114	2927	1.031	0.958	1113
J	12577	119	10599	2959	111	2667	0.876	1.012	986
S	12661	113	11250	2984	103	2900	1.026	0.963	995
D	12491	105	11974	3210	95	3380	1.079	0.997	1070
89:M	12471	102	12285	2939	88	3338	1.031	0.888	980
J	11911	95	12577	2399	81	2959	0.876	0.932	800
S	11164	88	12661	2237	75	2984	1.026	0.909	746
D	10201	82	12491	2247	70	3210	1.079	0.931	749
90:M	9143	73	12471	1881	64	2939	1.031	0.812	627
J	8184	69	11911	1440	60	2399	0.876	0.874	460
S	7356	66	11164	1409	63	2237	1.026	0.954	470
D	6794	67	10201	1685	75	2247	1.079	1.108	562

Technical planning studies

Akin to monitoring, the technical planning process involves both technical and market data. You search various sources in the literature for the current and emerging technologies in a specific area of interest. Then compare these with the technologies being used by the company. Results usually are more satisfactory, and more easily accomplished, when the subject area has been narrowed down—motor drive controls rather than control systems, fuel cells rather than power sources, and so forth. A task force doing several technical planning studies should discuss reports in brainstorming sessions or workshops and obtain possible threats or opportunities from the comparison of emerging and existing technologies. See Table 7.3 for an illustration of the process.

What is the impact of an emerging technology on your company?

Scenario formulation

A scenario is a narrative that takes us through time, say, the next 20 years, and presents an account of incidents or events that may occur in the public and private sectors as variables change under a particular theme (see Table 7.4).* Possible events may be new legislation, war, a cartel formation, and so on. Variables may be inflation rates, price increases, product demand, imports, competition, demographics, existing and new technologies, and annual cost increases. Themes may include the state of the economy, possible cartel formation, and various possible government actions, war, and a technological breakthrough. You realize technical forecasts by using the posed events, variables, and themes to obtain data on product supply and demand over the selected future period. By contemplating the effects of the forecast data in the scenario, you identify possible impacts on your technology and industry. These will be either threats or opportunities. Variations in themes (for example, a pessimistic or optimistic economy) may lead to different impacts and, hence, to variations in strategy, as discussed in Chap. 3.

Scenarios help us to formulate alternative futures.

Source material

In addition to the references given at the end of this chapter, Silk and Curley[138] provide a good listing of source material on forecasting. For a listing of sources for reading on statistics, see Wasserman.[139]

*Adapted from J. H. Vanston, Jr., et al., "Alternate Scenario Planning," *Techn. Forecasting & Soc. Change*, **10,** 159–180 (1977).

TABLE 7.3 The Process of Technical Planning

Monitoring	Inputs to assessment of study area	Initial assessment of study area	Technical forecasting
■ Surveillance over various areas of possible interest ■ Tracking of specific areas of interest from forecasting studies	■ Present literature (technical, legal, market) ■ Prior art ■ Company literature ■ Government reports ■ Competition ■ Expert opinion ■ Operations' inputs ■ Surveys (technical, legal, market)	■ Effects of emerging technologies ■ Threats and opportunities ■ Importance to company	■ Identify trends into future ■ Analyze technological and societal change ■ React to early warnings ■ Identify threats and opportunities ■ Estimate rate of takeover of a new technology

Technical Planning Report

- Sum up new technologies, threats, and opportunities that apply to company
  - Determine possible impacts
  - Provide recommendations

TABLE 7.4 Steps to Produce Scenarios

Define purpose and goals of the proposed scenarios.
Gather relevant data.
Identify, list, and select relevant variables (circumstances, facts, influences that produce a result).
Choose themes for alternative scenarios.
Group related variables.
Define the current situation, using values for selected variables.
Develop the most probable scenario, using various forecasting techniques.
Alter variables and prepare alternative scenarios.
Check all scenarios for consistency and reasonableness, and prepare final versions.

Decision Analysis

In R&D, decision analysis is a method of using uncertainty to arrive at economic evaluations of potential research. It aids in the selection of your R&D portfolio content,[140] and various companies using this technique attest to its usefulness.[141] In essence, you assess the sequential steps or crucial elements in the development of a new product or process, identify sources of uncertainty, and assign judgmental probabilities to obtaining favorable outcomes.

Decision analysis is valuable in evaluating and selecting research.

One decision analysis technique is the use of probability trees. For any event, probabilities of any outcome are between 0 and 1. The sum of the probabilities of all outcomes is 1, and with a series of possible outcomes, the probability that one *or* the other of any two will occur is the sum of their individual probabilities.

A probability (or decision) tree is then constructed in which the branches represent different scenarios that combine the development's various steps or technical possibilities. Analysis of the probability tree's scenarios provides the basis for decisions.

Decision analysis[142] allows for decisive action *before* and also *during* the project, and it focuses on the crucial aspects that determine uncertainty, thus making obvious the necessary R&D actions.

Before I get into some examples, you must understand the probability tree. This technique suggests that a value to be expected from a situation having several possible outcomes may be estimated by summing the results from the multiplication of the probability of success and the value for each outcome. This is shown below for two examples:

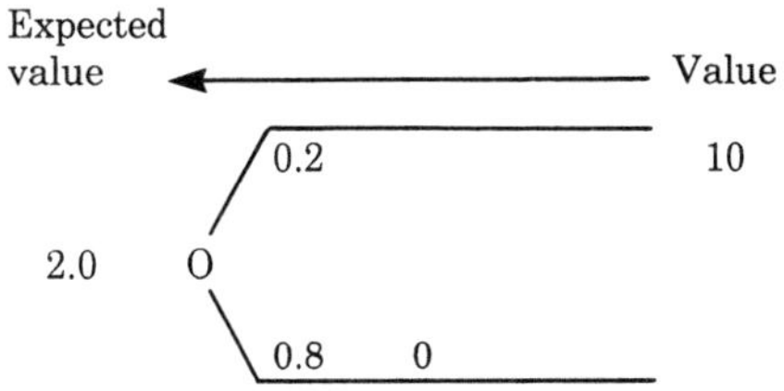

Expected value = (0.2)(10) + (0.8)(0) = 2.0

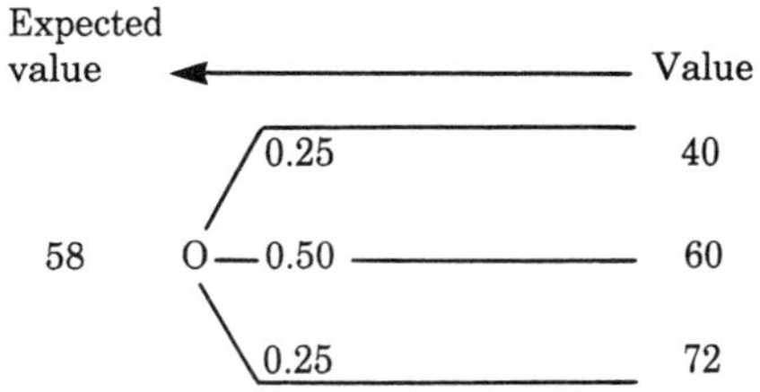

Expected value = (0.25)(40) + (0.5)(60) + (0.25)(72) = 58

In Fig. 7.4, the probabilities of solving certain technological problems depend on the research results. With a possible value of \$20 million and R&D costs of \$2 million, use of probabilities, as suggested be-

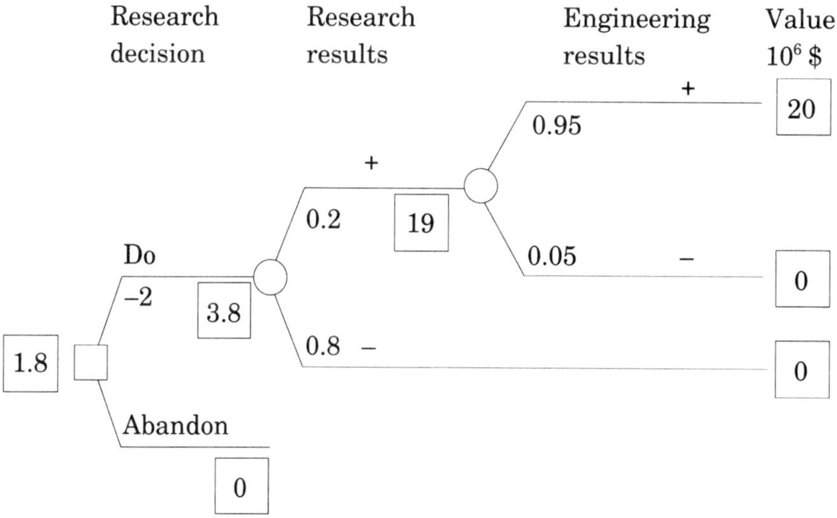

Fig. 7.4 Estimating the value of a project using a probability tree. *Note:* Overall probability of technical success = 0.19.

fore, leads to a positive expected value of \$1.8 million:

$$\{[(0.95 \times \$20 \text{ million}) = \$19 \text{ million}] \times 0.2 = \$3.8 \text{ million}\} - \$2 \text{ million (R\&D costs)} = \$1.8 \text{ million}$$

You can also use probability trees to decide whether to do R&D on one technology or on a combination of technologies, or to decide between two funding levels for a program.

Probability trees are also useful in illustrating to management the decision points or milestones, alternative routes to take depending on an outcome, and the values and probabilities associated with each alternative. Such a picture is not possible with a Gantt chart.

Another use of decision analysis is in evaluating a project portfolio.[143,144] Companies such as Exxon[145] have found no better method with which to evaluate their corporate research. In Table 7.5, I have evaluated 14 projects. Year of introduction, sales in the peak year, and probability of technical success were inputs. I obtained expected sales (ES) by multiplying peak sales and the probability of technical success P. I then made sales projections over the 10-year span. (Rather than personal guesstimates, these data would usually be obtained from a marketing group.) The summation of probabilities gave me an estimate of the number of successful projects, and the summation of expected sales gave an estimate of sales from the successful projects. The analysis doesn't tell me which projects will be successful. I have found that data typical of those illustrated in Tables 7.5 and 7.6 are very useful in the summary sections of annual reports and also for normal use by area managers.

To obtain the analyses in Table 7.5 (and 7.6), you need estimates of sales and probabilities of success. In the section on "Substitution Technology" in this chapter, I discuss how to get market share estimates. With knowledge of market share and total sales volume in the selected market segment, you get an estimate of your sales volume. That estimate multiplied by price gives the sales. Estimates of probabilities are more difficult. Here you need a panel of area managers who (with the help of the project leaders) estimate the probabilities of technical success for each project in the portfolio. The vice president of R&D reviews the estimates and makes a decision on the probability of technical success for each project. The original estimates and the vice president's values are returned to the panel which reviews all values and returns new estimates to the vice president. Again, decisions are made on the probability of success of each project, and these are the values used.

As refinements to this procedure, an interrogator can ask the panel for pessimistic, likely, and optimistic estimates on each project and then calculate for each input the probability of technical success using

TABLE 7.5 Evaluating a Project Portfolio by Decision Analysis

Project	Y	S	P	ES	1985	1986	1987	1988	1989	1990	1991	1992	1993	1994
Research area 1														
Project A	1983	10	0.8	8	1	1	2	4	8	8	8	8	8	8
Project B	1984	600	0.6	360	11	17	26	42	60	96	150	240	360	360
Project C	1990	5000	0.4	2000	—	—	—	—	—	120	180	275	340	450
Project D	1984	3	0.7	2	1	1	2	2	2	2	2	2	2	2
Research area 2														
Project E	1983	10	0.8	8	1	2	3	5	8	8	8	8	8	8
Project F	1983	60	0.8	48	2	4	9	21	48	48	48	48	48	48
Project G	1984	130	0.8	104	2	4	7	12	17	25	44	66	104	104
Project H	1984	600	0.7	420	5	8	15	25	43	78	136	231	300	420
Project I	1983	40	0.7	28	10	16	28	28	28	28	28	28	28	28
Research area 3														
Project J	1984	20	0.9	18	1	2	3	5	10	18	18	18	18	18
Project K	1984	100	0.8	30	3	6	11	21	40	80	80	80	80	80
Project L	1984	50	0.8	40	2	3	6	11	21	40	40	40	40	40
Project H	1986	400	0.7	230	—	3	6	12	26	58	120	280	280	280
Total			9.5	3396	39	67	118	188	311	609	862	1324	1616	1846

Note: Y = year of introduction; S = peak sales; P = probability of technical success; and ES = expected sales.

the method for estimating task times (see Chap. 6). Or the probability wheel[140] might be used to make the estimators think deeply about their answers. Again, the vice president of R&D decides on the values to be used. Such estimates should be made every six months to keep abreast of progress.

In Table 7.6, net present value (NPV) times probability of technical success (P) gives expected net present value (ENPV). The ENPV less R&D cost gives R&D value, and approximate R&D priorities are given by the R&D value divided by the R&D cost (with a higher index value indicating a higher priority). This is very much like some index methods of project evaluation such as:

$$\text{Index value} = \frac{P_{\text{TS}} \times \text{sales}}{\text{R\&D costs}}$$

where P_{TS} is the probability of technical success.

However, while such indexes are useful inputs to management's decision process, they do not tell the whole story. In particular, management should be aware that indexes are intrinsically biased against early-stage, low-probability projects. For this reason, Owen[143] of the Strategic Decisions Group recommends the two-dimensional display shown in Fig. 7.5. In this figure, "Bread & butter" projects have a high probability of technical success and generally have a low value. These may be product upgrades and are projects easily within the laboratory's capabilities. If the probability of success is misjudged and is lower than estimated, a "Turkey" can result. Projects with a high potential value and a high probability of technical success are the "Silver bullets." The "Question marks" have a low probability of technical success but have a high value if they succeed. This display is to help management appreciate the very different nature of different classes of projects (for example, bread and butter versus question marks) and thereby gain a greater awareness of the company's products, leading to a portfolio balanced with respect to time, risk, and return.

Be aware of the index bias.

Matrix Analysis

Although some people strongly favor matrix analysis, I prefer other methods for the R&D manager to master. While I will touch on this subject, I direct the interested reader to the reference section for more sophisticated and in-depth presentations.

The matrix technique is a tool used to tabulate, and make obvious valuable information that is hidden in a mass of data. In the "Strate-

TABLE 7.6 Evaluating Project Priorities by Decision Analysis

Project	NPV	P	ENPV	R&D cost	R&D value	R&D index
Research area 1						
Project A	24	0.1	2	1	1	1.0
Project B	492	0.1	49	25	24	1.0
Project C	449	0.4	180	25	155	6.2
Project D	9	0.7	6	1	5	5.0
Research area 2						
Project E	26	0.8	21	2	19	9.5
Project F	134	0.1	13	7	6	0.9
Project G	138	0.1	14	20	-6	- 0.3
Project H	435	0.4	174	20	154	7.7
Project I	121	0.1	12	15	-3	- 0.2
Research area 3						
Project J	45	0.1	41	3	38	12.7
Project K	191	0.3	57	9	48	5.3
Project L	97	0.2	19	5	14	2.8
Project M	362	0.4	145	18	127	7.1

Note: NPV = net present value; P = probability of technical success; and ENPV = expected net present value.

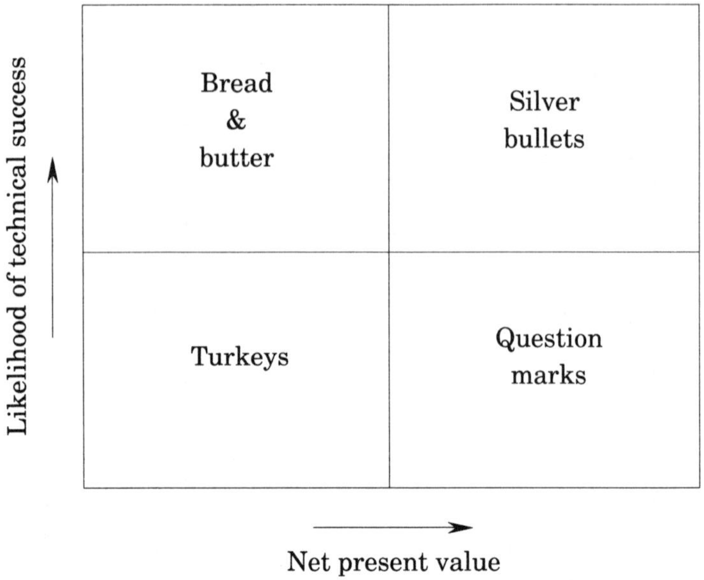

Fig. 7.5 Selecting a balanced project portfolio.

gic Technical Planning" section (Chap. 3), I discussed information gathering and consolidation stages, and the matrix techniques can be an aid in these areas. For example, a matrix comparing product lines versus basic markets shows a product's current status in the market. Markets being serviced and the completeness of product lines become obvious. Or a competitive product matrix can be made of product versus company of manufacture. When considering where your company might want to go, you can prepare a matrix of products versus companies that could be candidates for acquisition. Matrix analysis is also an aid in screening the question of whether to obtain a product through internal R&D or by acquisition. New product opportunities are highlighted by obvious voids in product lines shown on a matrix. You can also use matrix analysis in market segmentation to discover opportunities.

In portfolio analysis and strategy formulation, many types of matrices are used; however, I have selected four for some examination: the growth/share, the business profile, the business assessment, and the technology portfolio matrices.

The Boston Consulting Group[146] produces the growth/share matrix (Fig. 7.6) for a portfolio of products. A series of such charts over time indicates each product's direction and rate of movement. Strategies are then developed on the basis of where the products are and their movement in comparison with similar charting of competing products.

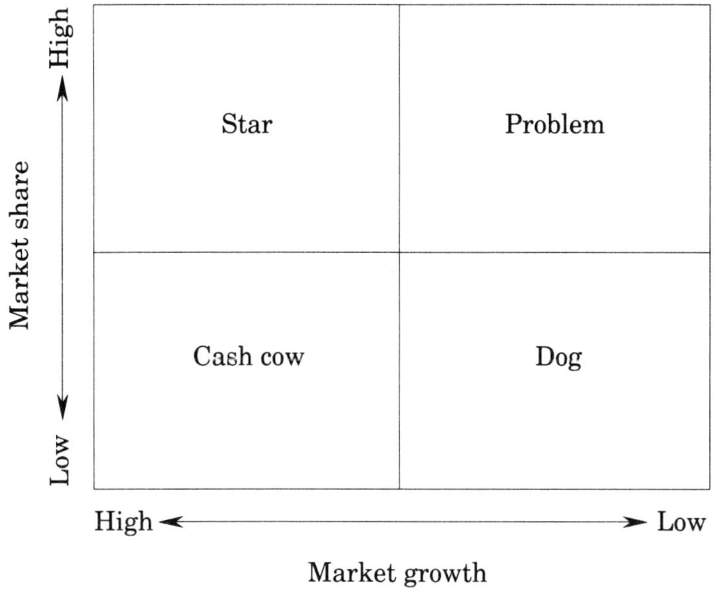

Fig. 7.6 Growth/share matrix. (*Reprinted by permission, from* Perspectives, *no. 66, "The Product Portfolio," copyright ©1970, The Boston Consulting Group, Boston, Mass.*)

The business assessment matrix of General Electric[147] (Fig. 7.7) stresses that investments in future growth should be derived from an appropriate mix of business strengths and industry attractiveness. (If you use product/market attractiveness versus company capabilities, you have the PIMS 3 × 3 cell matrix.[45])

Arthur D. Little[148] believes that the strategic condition of a business depends upon the maturity of the industry and its competitive position within that industry. Identifying various businesses on a business profile matrix (Fig. 7.8) permits visual comparison of one business with another. Analyses of products, technologies, and possible future change then suggest how each business will grow. On this basis, business strategies are evolved.

Booz-Allen and Hamilton[49] compares business and technology portfolios in 2 × 2 cells (Fig. 7.9) to derive conclusions on better technologies to support.

Technology Futures has developed a management approach called "Innovation by Design" which uses a matrix called the "applied innovation map" (AIM). As seen in Fig. 7.10, innovations are categorized into one or more of the cells and the matrix is then put through a series of analyses. The result is said to lead to an effective innovation program.

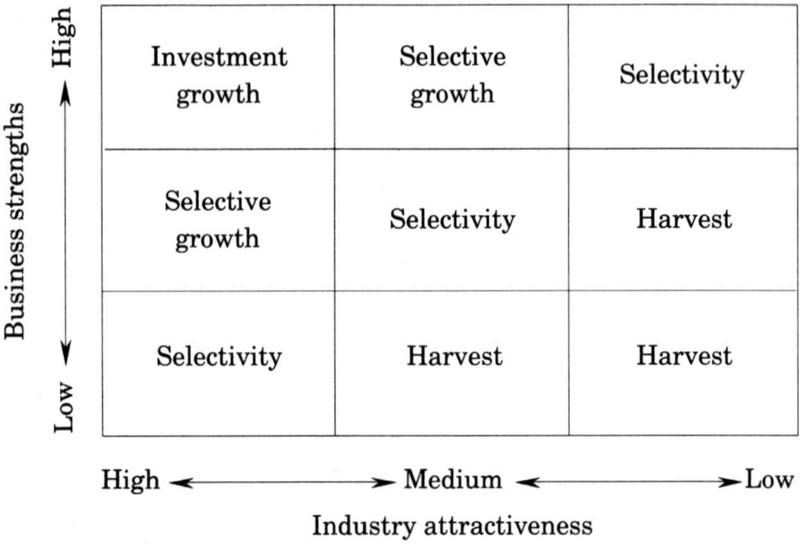

Fig. 7.7 Business assessment matrix. (*Reprinted by permission of the publisher, from* Corporate Planning, *by Robert J. Allie and Malcolm W. Pennington © 1979 AMACOM, a division of American Management Association, New York. All rights reserved.*

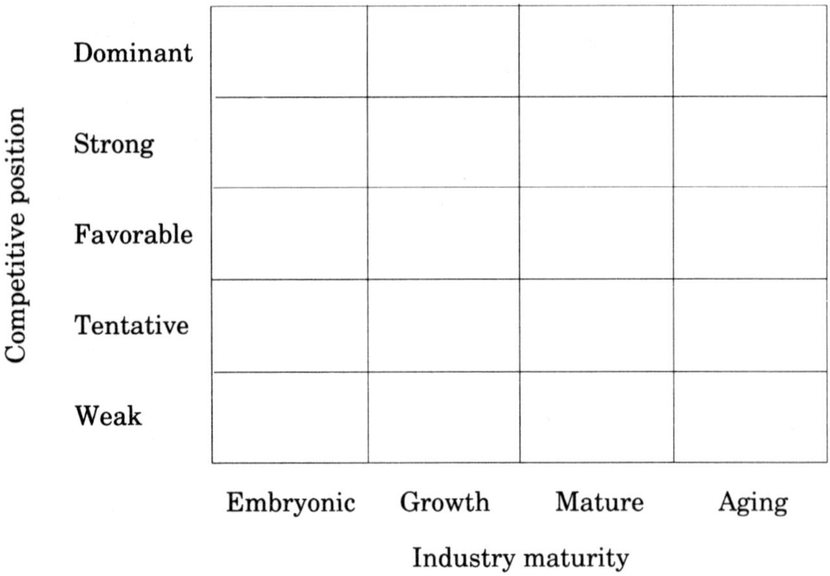

Fig. 7.8 Business profile matrix.

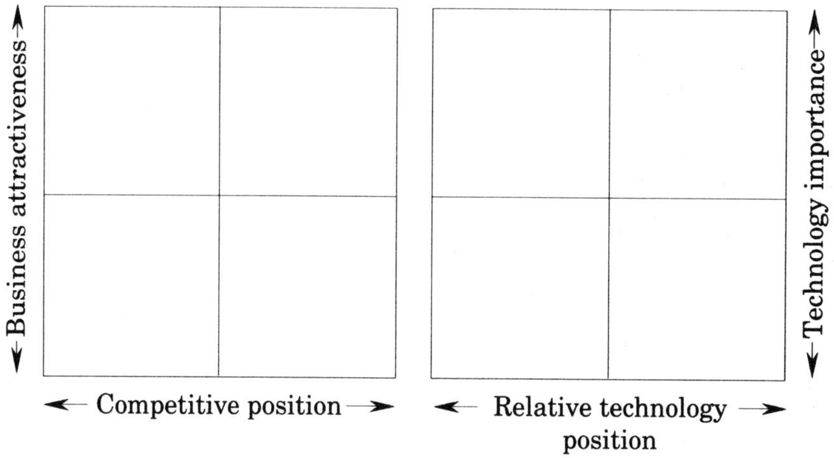

Fig. 7.9 Technology portfolio matrix. *Source: Booz • Allen & Hamilton, Inc.*

Class

Nature		Incremental	Distinctive	Breakthrough
	Product			
	Process			
	Procedure			

Fig. 7.10 Applied innovation map.

Summary

- Forecasting can be *one* of the very handy tools you use in planning.
- Forecasting is not an exact science, and forecasts must be systematically monitored.
- Forecasts using one technique should be compared with forecasts using another technique to firm up your estimate.
- Technical planning studies are useful in assessing how emerging technologies may affect your company's future.
- Scenarios are narratives of alternative futures.
- Decision analysis is a method using uncertainty to obtain evaluations of potential research and to aid in formulating project portfolios.
- Probability trees illustrate decision points and aid in decisions on alternative routes of R&D.
- Matrix analysis is useful in making valuable information obvious from a mass of data.

Cases and Incidents

Giving in Order to Get

Shock finds that reading the division's annual reports gives him an idea of products sold and projects underway but does not disclose the division's problems. He also knows that the best way to get a handle on these problems is to make one or more visits to divisions' managers, depending on how close to the vest division personnel are holding their cards.

Shock tells Watts that when someone finally says, "We've got this problem," he knows a communications breakthrough has been made and they are on their way to some constructive rapport.

Watts wants to know how Shock gains this rapport.

Shock replies, "It's not easy. Everyone distrusts an outsider. And because we are from a corporate group means (to them) that we are in an ivory tower and don't know real-life needs. Also, they don't want the wrong interpretations going upstairs. But I have this routine that seems to work. I describe one or two problems that they might help with, and venture some guesses as to where our R&D might help them. Our problems show that we are human. Also, I have reviewed their products, so I am well prepared to make some legitimate guesses about how our efforts can help theirs. You can see a warming as I go through each part of my presentation, which I try to make very informal. The conversation gradually gets more casual and friendly. As I said, if I'm successful, they always say 'I've got this problem' and launch into some guesses of their own on how our work might be an answer."

Watts congratulates Shock, "Your technique is great. You recognize that you have to give information to get information. And the more you learn about operations, the better you can develop your research programs."

Shock smiles. A compliment is a real boost, even to a vice president of R&D.

Questions

- How would you gain rapport?
- Is more than one meeting necessary—for example, the first by Shock to get acquainted and the second by a selected professional to get into specific technical matters in detail?
- Or could Shock visit and have visits from division personnel?

The Kiss of Death

Larry, manager of the applications development area, has an idea for a new pump which ensures against cavitation and promises greater efficiency. He discusses it with the local offices of the pump division. The division's local chief engineer, Delaney, and Larry hit it off well and agree that it would be good to build a prototype and test it.

A proposal by Larry is forwarded to the division president by Delaney. The president turns the proposal over to his division planner for review. The following memorandum is received by the local division office manager from the division planner:

> Noted is Delaney's recent memo to Bob. Let me save you some time. About twenty years ago, I tested such a pump and found its efficiency extremely low. Complex design led to impossible tolerances that could be achieved only in a lab prototype, and downtime to replace components was exorbitant. This pump is not new to the art, and nothing will be gained by the proposed test.

It is clear to R&D that the planner has failed to grasp the innovative qualities of the new pump.

Shock asks, "What the hell do we do now?" Watts says that the fact that the division planner missed the subtleties of the design will not offset the damage done by his memo. Had Delaney walked the proposal through, the new aspects of the pump could have been described and any comments discussed. The planner would have been able to recognize the unique merits of the new pump design.

Watts suggests that Shock phone the planner and tell him he is absolutely right about the pump he describes but that the proposal concerns a pump with a different design and, unfortunately, this was not adequately expressed in the proposal. Then Shock should say that he is asking Delaney and Larry to rewrite their proposal, make the design differences clear, and that Delaney should take it to the planner and the president and defend it.

Shock says, "Walk it through, you mean."

"That's right," Watts replies.

Questions

- Should the planner's memo be answered? If so, by whom and when?
- Should Delaney have known about the planner's background and checked with him earlier?
- If the planner can't be convinced of the differences, will the project be undertaken? If it is undertaken, will it be successful?

References

131. T. Gordon and O. Helmer, "Report on a Long-Range Forecasting Study," Rand Pap. P-2982, 1964. This reference is given in H. A. Linstone and M. Turoff, *The Delphi Method,* Addison-Wesley, Reading, Mass., 1975.
132. R. C. Lenz, "Forecasting the Rate of Adoption of New Technology," July 1980 (text for a course given by Technology Futures, Inc., Austin, Tex.). See also J. C. Fisher and E. H. Pry, "A Simple Substitution Model of Technological Change," *Techn. Forecasting & Social Change* **3**:75–88 (1971).
133. A. W. Blackman, "An Innovation Index Based on Factor Analysis," *Techn. Forecasting and Social Change* **4**:301–316 (1973) and "The Market Dynamics of Technological Substitutions," *Techn. Forecasting and Social Change* **6**: 41–63 (1974).
134. W. E. Souder and A. Quaddus, "A Decision-Modeling Approach to Forecasting the Diffusion of Longwall Mining Technologies," *Techn. Forecasting & Social Change,* **21**:1–14 (1982).
135. J. W. DeWolf, J. Neth, and K. Gruszcynski, *How to Make Early Warning Forecasts,* 3d ed., Cahners, Boston, 1974, and *Early Warning Forecast,* 13th ed., Cahners, Boston, 1983.
136. W. L. Schubert, "Forecasting Drilling Activity," *Oil & Gas J.,* Nov. 15, 1982, pp. 104–110.
137. N. Levenbach and J. P. Cleary, *The Beginning Forecaster: The Forecasting Process through Data Analysis,* Lifetime Learning Publications, Belmont, Calif., 1981.
138. L. S. Silk and M. C. Curley, *A Primer on Business Forecasting,* Random House, New York, 1970.

139. P. Wasserman (ed.), *Statistics Sources,* Gale Research, Detroit, Mich., 1983.
140. H. U. Balthasar, R. A. A. Boschi, and M. M. Menke, "Calling the Shots in R&D," *Harvard Bus. Rev.,* May–June 1978, pp. 151–160.
141. Technical Insights, Inc., "Using Decision Analysis to Determine R&D's Value," *Inside R&D,* March–April 1983.
142. A. J. Pacey and G. P. Murray, "Climbing the Decision Tree," *Chemtech,* March 1984, pp. 156–161.
143. D. Owen, "Selecting Projects to Obtain a Balanced Research Portfolio," Pap. A83–33, Metallurgical Society of AIME, Warrendal, Pa., March 1983.
144. Technical Insights, Inc., "Managing Your R&D Portfolio," *Inside R&D,* June 1984.
145. Technical Insights, Inc., "How Exxon Evaluates Its Corporate Research," *Inside R&D,* April 1984.
146. B. D. Henderson, "The Experience Curve Reviewed. IV: The Growth Share Matrix or the Product Portfolio," Pap. 135, Boston Consulting Group, Boston, 1970.
147. M. G. Allen, "Diagramming GE's Planning for What's WATT," in R. J. Allio and M. W. Pennington, *Corporate Planning: Techniques and Applications,* AMACOM, New York, 1979.
148. "Strategy Formulation in an Uncertain Future" and "Strategic Management of Technology," Arthur D. Little, Cambridge, Mass., 1981.

Suggested Readings

Armstrong, S. S.: *Long-Range Forecasting,* Wiley, New York, 1978.

Bright, J. R.: *Practical Technology Forecasting,* Communication Specialists, Inc., Austin, Tex., 1980.

Day, G. S.: "A Strategic Perspective on Product Planning," in *Readings in the Management of Innovation,* M. L. Tushman and W. L. Moore, Pitman, Boston, 1982.

Hopkins, D. S.: "New Emphasis in Product Planning and Strategy Development," *Ind. Mark. Mgmt.,* **6**:410–419 (1977).

Jackson, B.: "Decision Methods for Evaluating R&D Projects," *Res. Mgmt.,* July–August 1983, pp. 16–22.

Makridakis, S., and S. C. Wheelwright: *The Handbook of Forecasting,* Wiley, New York, 1982; see especially chap. 6, "Forecasting in Strategic Planning," by P. Lassere and H. T. Thanheiser.

Mansfield, E.: "Technical Change and the Rate of Imitation," *Econometrica,* **29**(4):741–765 (1961).

Mitchell, A., et al.: *Handbook of Forecasting Techniques,* SRI International, Menlo Park, Calif., December 1975.

Raiffa, R.: *Decision Analysis,* Addison-Wesley, Menlo Park, Calif., 1970.

Schlaifer, R.: *Analysis of Decisions under Uncertainty,* McGraw-Hill, New York, 1969.

Sheldon, R. A.: "How Forecasters Forecast," *Chemtech,* April 1978, pp. 204–208.

The Conference Board: "Generating New-Product Ideas," New York, 1972.

Vanston, J. H., Jr.: "Technology Forecasting: An Aid to Effective Technology Management," Technology Futures, Inc., Austin, Tex., 1982.

Wind, Y., and V. Mahajan: "Designing Product and Business Portfolios," *Harvard Bus. Rev.,* January–February 1981, pp. 155–165.

Special Bibliography on Technological Substitution

Ayres, R. U.: "Technological Forecasting, An Economic Approach," Dept. Eng. and Public Policy, Carnegie-Mellon University, Pittsburgh, Pa., November 1983.

———: "A Schumpeterian Model of Technological Substitution," *Techn. Forecasting & Social Change,* **24**(4):375–383 (1985).

Bass, M.: "A New Product Growth for Model Consumer Durables," *Management Science,* **15**(5):215–227 (1969).

Blackman, A. W., Jr.: "The Rate of Innovation in the Commercial Aircraft Jet Engine Market," *Techn. Forecasting & Social Change,* **2**:269–276 (1971).

———: "A Mathematical Model for Trend Forecasts," *Techn. Forecasting & Social Change,* **3**:441–452 (1972).

———: "New Venture Planning: The Roles of Technological Forecasting," *Techn. Forecasting & Social Change,* **5**:25–49 (1973).

———: "The Market Dynamics of Technological Substitutions," *Techn. Forecasting & Social Change,* **6**:41–63 (1974).

———, E. J. Sligman, and G. C. Sogliero: "An Innovation Index Based on Factor Analysis," *Techn. Forecasting & Social Change,* **4**:301–316 (1973).

Bretschneider, S. I., and V. Mahajan: "Adaptive Technological Substitution Models," *Techn. Forecasting & Social Change,* **18**(2):129–139 (1980).

Dowling, G. R.: "Application of the Bass New Product Growth Model to Australian Television Sales," *Australian J. Mgmt.,* **5**(1, 2):133–140 (1980).

Easingwood, C., V. Mahajan, and E. Muller: "A Nonsymmetric Responding Logistic Model for Forecasting Technological Substitution," *Techn. Forecasting & Social Change,* **20**(3):199–213 (1981).

Haines, G. H., Jr.: "A Theory of Market Behavior after Innovation," *Mgmt. Sci.,* **10**(4):634–649 (1964).

Hurter, P., Jr., A. H. Rubenstein, M. Bergman, J. Garvey, and J. Matinich: "Market Penetration by New Innovations: The Technological Literature," *Techn. Forecasting & Social Change,* **11**:197–221 (1978).

Jackson, J. R., and D. L. Kaserman: "Neoclassical Theory of Durable Goods Diffusion," Oak Ridge National Laboratory, NTIS, PC A05/MF A01, October 1979.

Kabir, C., M. N. Sharif, and P. Abdulbhan: "System Dynamics Modeling for Forecasting Technological Substitution," *Computers & Ind. Eng.,* **5**(1):7–21 (1981).

Lee, J. C., and K. W. Lu: "On a Family of Data-Based Transformed Models Useful in Forecasting Technological Substitution," *Techn. Forecasting & Social Change,* **31**(4):61–78 (1987).

Leoparelli, L., and M. Lucertini: "Substitution Models for Technology-Fostered New Production Inputs," *Techn. Forecasting & Social Change,* **16**(2):119–142 (1980).

Linstone, H. A., and D. Sahal: *Technological Substitution,* Elsevier, New York, 1976.

Mahajan, V., and E. Muller: "Innovation Diffusion and New Product Growth Models in Marketing," *J. Marketing,* **43**(Fall):55–68 (1979).

Mansfield, E.: "Technical Change and the Rate of Imitation," *Econometrica,* **29**(4):741–766 (1961).

Nevers, V.: "Extensions of a New Product Growth Model," *Sloan Management Review,* Winter 1972.

Philipson, L. L.: "Market Penetration Models for Energy Production Devices and Conservation Techniques," *Techn. Forecasting & Social Change,* **11**:223–236 (1978).

Sharif, M. N., and A. K. M. Haq: "Determinants for Forecasting Technological Substitution," *Techn. Forecasting & Social Change,* **13**(1):59–81 (1979).

———, and C. Kabir: "A Generalized Model for Forecasting Technological Substitution," *Techn. Forecasting & Social Change,* **8**:353–364 (1976).

———, and G. A. Uddin: "A Procedure for Adapting Technological Forecasting Models," *Techn. Forecasting & Social Change,* **7**:99–106 (1975).

Silvennoinen, P., and J. Vaananen: "Forecasting Technological Substitution: The Logistic Model of Energy Systems Revisited," *Techn. Forecasting & Social Change,* **32**(3):273–280 (1987).

Souder, W. E., and A. Quaddus: "Decision-Modeling Approach to Forecasting the Diffusion of Longwall Mining Technologies," *Techn. Forecasting & Social Change,* **21**(1):1–14 (1982).

Stern, M. O., R. V. Ayres, and A. Shapanka: "A Model for Forecasting the Substitution of One Technology for Another," *Techn. Forecasting & Social Change,* **7**:57–79 (1979).

Tigert, D., and B. Farivar: "The Bass New Product Growth Model: A Sensitivity Analysis for a High Technology Product," *J. Marketing,* **45** Fall:81–90 (1981).

Chapter

8

Techniques for Implementation Activities

Questions answered in Chapter 8

- Why is it difficult to make significant advances with a mature technology?
- We need various screening and evaluation methods. Why?
- How do we establish an idea's feasibility?
- How is decision analysis used in portfolio analysis?
- What are the elements of market research?
- If patents are necessary, why must they be controlled?
- In a licensing negotiation, why not try to get everything you can?

To carry out the strategic plan, you must have many ideas, cull the poorer ones, nurture the good ones, develop them, and pass the developments on while being creative, efficient, cooperative, and communicative.

You must initiate research with only the best ideas; to get the best, you must have many and select the few worthy of development, taking care not to destroy creativity in their evaluation. There are a great many ways to evaluate ideas, and I present below a few that I believe should receive your consideration. References are given, should you desire to investigate others.

Ideate, screen, and go only with the best ideas.

Comparisons of actual and expected accomplishments and benefits versus costs enter into periodic project reviews and project portfolio reviews. Again, I advance some favored methods for your consideration. Decision analysis, for instance, is important to this discussion.

Market research is a necessary adjunct to your technical work, and you may wish to farm much of this out. In so doing, you need to be aware of the elements of market research, and I present these along with some practical observations.

Finally, you, as managers, must have an understanding of patents and licensing, since most companies are highly interested in protecting proprietary information. However, you must keep control of your professionals' zeal to patent because of costs. An unused patent is wasted money and wasted energy.

Evaluating Ideas, Projects, and Project Portfolios

Fundamental to the idea or project to be evaluated is the technology it uses. Thus, you should know when current technology is diminishing in value and other technologies are advancing. Foster[94–96] suggests that a technology's position on the S-curve (see Fig. 7.1 and related discussion) determines the potential that remains in that technology, and that selecting the right technology should be given first priority in the development of a project portfolio. He also suggests that no technology is limitless and that R&D productivity can be considered as the slope of the S-curve, because it represents the amount of progress provided by a given input of effort. Also see Becker[149] and Pogany.[150]

Design your screening or evaluation technique to fit the stage of development.

As stated in the Preface and in Chap. 7, as a technology matures, R&D productivity using that technology declines and it is harder to make any substantive innovative advance. You should note a decline in technological progress for a given input of effort: the progress of technology along the S-curve begins to deviate from its exponential growth. Thus, a technology's position on the S-curve determines the potential remaining to be developed. Roussel[151] illustrates the technology maturity cycle by plotting the number of significant publications on a given technology versus time. As maturity approaches, increasing attention is given to process changes rather than to product changes in an effort to reduce costs. You must recognize when one of your technologies is aging and becoming susceptible to displacement. This is the time to look for, and adopt, a new technology, resulting in a new S-curve (see Fig. 8.1).

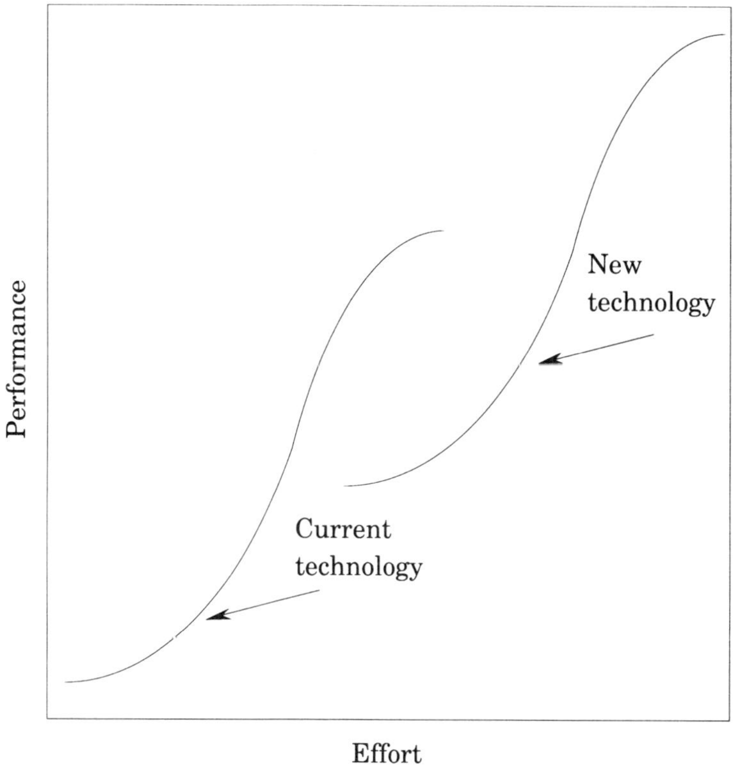

Fig. 8.1 New technology.

Is there potential for further development in your products?

Screening and evaluation techniques should be designed to recognize the six phases of the innovation process and the fact that innovation starts with a new, rough idea. Instead of this approach, many techniques explore financial or risk aspects of products that have already had some R&D exploration, a task that requires knowledge of the product's characteristics and markets. Recognition of the phases of innovation permits use of more relevant evaluation procedures.

In his evaluation procedures, Albala[152] uses checklists (listings of characteristics) for exploratory work (feasibility study), profile charts (graphical evaluation of relevant criteria) or scoring models (numerical evaluation of relevant criteria) for applied research, and economic criteria (profitability and probability of success) for development projects. (In his text, Albala gives a great many references for each type of procedure.)

I like scoring models[153] for idea and project screening (see next section). The graded criteria change with the stage of evolution of the project; however, an evaluation of possible technical, legal, and market constraints must be made at all stages. In portfolio evaluation, I like to use decision analysis.

Idea screening

Documenting a new idea is essential to record its birth, give it formal existence, and make it available for idea review. Professional staff should initially document an idea in a bound notebook, either the project notebook or one kept for general use. Later, they should record it on an idea record form (see Chap. 6, Exhibit 6.1) and file it appropriately until an idea review committee meets to evaluate the accumulated ideas and recommend feasibility studies. Five people representing various disciplines and experiences should be on the idea review committee.

The idea review matrix shown in Exhibit 8.1 allows you to evaluate an idea using six criteria (the number is arbitrary and can be changed as you see fit). Any score of less than 20, or a rating of 1, would lead to rejection in this instance. If the average rating from the five-member committee is in the 25–30 rating range, the committee recommends a feasibility study. If the rating is 20–25, the committee asks the originator of the idea for additional detail. (The originator may be in the meeting during the review and participate as needed.) Similar matrices can be developed for the later stages of innovation; in such cases, more detailed questions are asked and more quantitative data are required. Guidelines that reflect the specific markets and products of each business are needed for scoring in the latter case. Use the experience of colleagues and consultants and the reports in the literature to assess these. Assessing constraints by whatever method used also allows you to estimate the project's probability of technical success.

Project evaluation

If the feasibility of an idea is established through investigation of technical, legal, and market constraints, a project is proposed based on a more in-depth investigation of constraints and the development of a prototype. Any of the evaluation methods (checklists, scoring models, and economic criteria) used to evaluate the idea can be expanded to evaluate the project, with any criteria added to a scoring matrix re-

Title ______________________

Reviewer ______________________

Date ______________

Score ______________

IDEA REVIEW

Criterion \ Rating	Unfavorable		Neutral	Favorable	
	1	2	3	4	5
Technical soundness					
Market potential					
Novelty					
Company relevance					
Competitive advantage					
Constraints and degree of risk					
Total					

25 30 Rocommond for feasibility study
20–25 Ask for more detail
20 (or with any rating of 1) Reject

Reasons for Unfavorable Ratings

How Constraints Might Be Overcome

Exhibit 8.1 Sample idea review form.

quiring more quantitative answers than required during idea screening.

Pacifico and Polacek[154] found that 50 percent of short- and medium-range projects were successful when selected using the *project index number* (PIN) procedure. This procedure is based on the following equation:

$$\text{PIN} = \text{risk}\,\frac{\text{total benefits}}{\text{total investment}} = R_t R_c R_l R_o \frac{A}{B}$$

where $A = V(P - C)L + \text{SV}$

$B = I_{mr} + I_{lr} + I_{pc} + \cdots I_o$

R = probabilities of success, including technical (t), commercial (c), legal (l), and other (o)

I = investments in marketing research (mr), laboratory research (lr), patents (pc), development (md), plant and equipment (pe), working capital (w), startup losses (sl), and other (o)

L = the expected product life in years

V = the annual sales in units

P = the selling price in dollars

C = the unit cost in dollars

SV = the salvage value

Merrifield[155] advances another project evaluation method. His business planning score chart is based on criteria involving business attractiveness and company strength, both of which are fundamental to commercial success. The planning score chart is then translated to an identity chart which visually displays the projects competing for funds and depicts the commercial probability of success as contrasted with the technical probability of success (refer to the discussion of decision analysis in Chap. 7). The multiplication of both probabilities gives an index of merit.

You could use Merrifield's method in the strategy formulation stage of your strategic planning (Chap. 3). One caution, as suggested in that chapter, concerns giving total reliance to the outcome of such analysis, especially if you are getting involved in a new technology in which the company fit could show weaknesses in marketing, manufacturing, or technology. Of course, the company could arrive at a strategy to overcome these weaknesses and use the new technology profitably. In any case, the method provides a valuable checklist for broadly assessing risks and benefits. Increasing costs of development over time makes the first go or no-go decision extremely important (see Fig. 8.2).

You can also use Merrifield's method in project evaluation to check the adequacy of business attractiveness and company strengths, and use Pacifico and Polacek's method to rank projects according to relative benefits under varying probabilities.

Portfolio management

Matrix analysis is useful in portfolio analysis. Goldgehn and Lagarce[156] combine portfolio analysis (such as that obtained by using

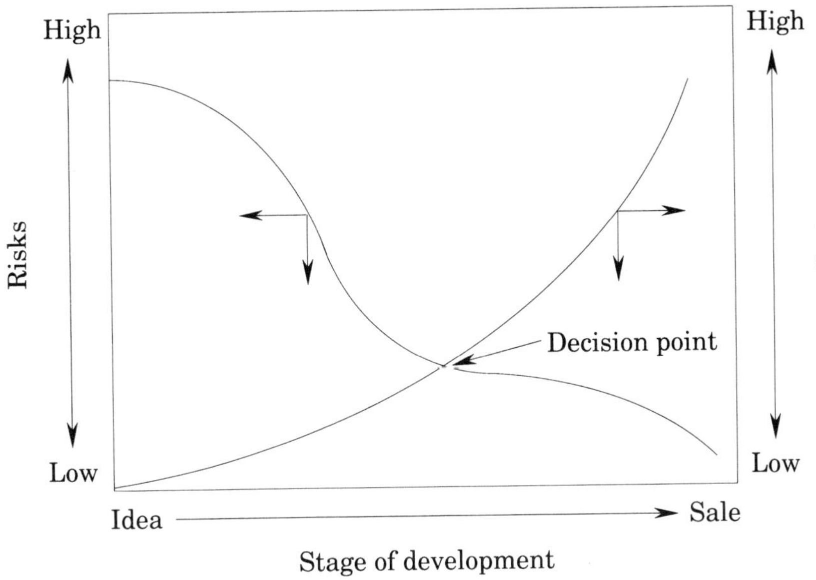

Fig. 8.2 Risks and costs versus stage of development. *Source:* AMA. Management Briefing, *AMACOM, New York, 1977, p. 8.*

the Boston Consulting Group's matrix) with product life cycle analysis in order to make adequate strategic decisions.

I prefer to use decision analysis to rank projects and help shape the portfolio. Tabulated data over a period of time (such as given in Tables 7.5 and 7.6, Chap. 7) are used to chart a project's progress and to check the principal investigator's progress. For example, such data could point to an increasing probability of technical success over time, a modification of sales figures as a result of the requirement for revised market studies, a modification of proposed year of introduction as a result of revision based on current review of technical progress, and a change in the R&D index over time.

In portfolio evaluation, you must moderate the findings of decision analysis. Always include consideration of the "right" proportion of short- and long-term projects. If the general economy is poor, give increased emphasis to the short term but never exclude the long term. Also, you must have some short-term projects that are expected to pay out while the longer-term project is under way (see Fig. 8.3). Ideally, this balance is one that R&D and top management should mutually decide. Questions that must be answered pertain to the charter and goals given by top management and to top management's views about company needs and risk attitudes. For example, what are top manage-

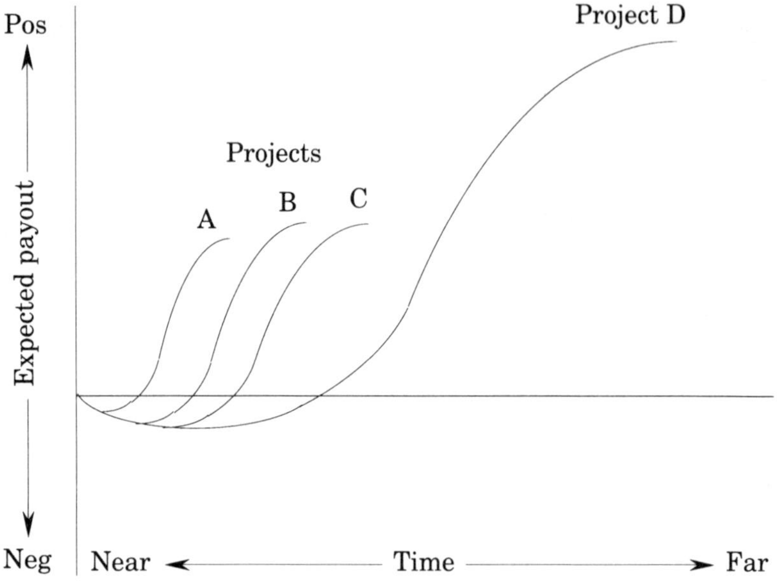

Fig. 8.3 Payout balance among projects.

ment's views on the proportion of new product R&D that has company relevance? on new product R&D in entirely new business areas? on problem solving for existing products? and so on.

Market Research

The following is based, in part, on a speech by E. D. Porro.* R&D people should have an understanding of market research and the problems that a poorly conducted research study can pose. Within a central research group, the staff (or an outside agency) accomplish preliminary market studies to qualify ideas for study and to aid in convincing an operating division of an idea's value. During project work, an operating division, an outside market research group, or a consultant would probably perform the market research.

The critical importance of market research to successful innovation has been confirmed in many studies.[97,155,157,158] Failure to assess a market accurately is due to misreading customer needs, not spotting early signs of competitors' offensive actions, performing too little field testing, and making overly optimistic forecasts of market need and acceptance. Market research done early in a project may become outdated and useless by not keeping up with new internal and external

*E. D. Porro, "Roles of Market Research," 3d annual Engineering and Technical Conference, Dresser Industries, Dallas, TX, November 13, 1978.

findings. Problems from outside market research may also be traceable to you, the buyer of this market research. You must ask the right questions; you and the contract research house must be on the same wavelength; and you must be willing to pay for the scope of research desired. You must, of course, choose a research firm or consultant that has competence and experience fitting the area of interest. You don't want reworked older market research.

Ask the right questions of market researchers.

Quick and dirty early in-house market research on a new idea may be inadequate or downright fallacious. Preconceived ideas of market characteristics (such as size) and of market need may dominate early discussions, and it is important that early market input be objective and factual. There is the story about a marketing group telling an R&D group that a $100 million market existed if widget A could be developed. When R&D presented engineering with a widget A prototype, engineering had a market study done which showed only a $1 million plus market. What happened? R&D had blindly accepted the first market estimate and had made no attempt to verify it. To avoid failure within the marketing constraint,[158] R&D must not only understand the problems of, and get along with, the various marketing groups within its company but also validate any market data it receives.

If you have a new product and want a market research study to determine whether it can compete in the marketplace and yield a reasonable profit and market share in a reasonable time, there are two stages of market research to consider: whether the market is satisfactory and whether your product will sell into it and return an adequate profit.

In the first stage, the determination of whether the market is satisfactory, you will investigate, or have investigated for you, such factors as whether a real need exists for your product and customers appear willing to buy. A need may exist, but because of price or some other reason, the willingness to buy is nonexistent or minor. There is always the possibility that someone in the company will merely propose that a customer will buy the new product, without actually determining objectively if this assumption is true. (Research people are generally accused of such a position, but it can be catastrophic if a top management person takes this position and ignores objective evaluation of the market data.)

If there is a need and a willingness to buy, you will then determine, as part of the first stage, the market structure, which includes identification of:

- The characteristics of the market, including size and growth. In particular, you must obtain statistics on the *segment* of the market into which your product may sell. Data on the *total* market can be misleading.
- The characteristics of the customers, such as whether they are locked into a specific product or shop for price, quality, or both.
- The product specifications required by the customer. Existing products that yours will replace will lead to these specifications; your product should exceed existing performance specifications.
- Any restrictions that may affect the sale of the product; for example, local air pollution regulations.
- The nature and extent of competition, such as the number of competitors, whether there is a dominant one, and the strength of product loyalty.

If the data from the first stage do not meet the market characteristics required by your company, work on the product should be terminated. If the data are favorable, you will determine in the second stage whether the product will sell into the market at a reasonable profit, gaining a reasonable market share in a reasonable time. The inputs needed include:

- Data on the new product's performance. Its performance must exceed the specifications obtained in the first stage.
- The comparative advantages of the new product and the probable investment the customer must incur on its purchase. What rate of return can the customer expect with your product versus the competitor's? What does it cost the customer to switch products? Will it be difficult to overcome product loyalty? Will the benefits relative to costs be in your favor?
- The appropriate timing for a future market entry. Does this time match with estimated product development time? If you don't meet appropriate deadlines, how likely is it that a competitor could use this time to saturate the market?
- Distribution channels and requirements. Is there relevant company experience? If not, can this experience be acquired? Someone must know how to *sell* the product!
- Price is, of course, an important factor in estimating sales. You also need to estimate how your product introduction may influence competitors' pricing and how that, in turn, may affect achieving a satisfactory return on investment.
- The estimated share of the market that can be captured. Make the

estimate from interpretation of data collected on relative product advantages, market characteristics, and pricing by you and competition. Use adjunct methods such as Fisher-Pry and Blackman's (Chap. 7) and Pacifico and Polecek's (Chap. 10).

- Will the uniqueness of the product lead to a patent structure that will ward off encroachment by competition?

If the data from the second stage do not meet the company's market criteria, then scrub the project; otherwise, proceed.

Both of these stages can be done with varying degrees of sophistication or completeness—from a quick and dirty fix to an in-depth study. These generally match with the maturity of the concept, becoming more complete as development proceeds.

At some point, the technical group must decide which market segment to enter first and then tailor the product to that market. This may not be the largest market, but it is a market in which a beachhead can be secured with the least investment gaining a good market share in the shortest time. Such a market has the greatest probability of commercial success; it is a judgment call. Market research should pinpoint this first market entry and its characteristics.

Market researchers collect data by several means and from several sources. Personal interviews are the main method because discussions can ensue about complex products and because the researchers can vary their interviews depending on the respondents' comprehension. Telephone interviews are useful if the product is well defined and easily describable, but this method is best for follow-up to supplement personal interviews. Success is a function, in part, of the number of the researchers' acquaintances. Inputs will also be obtained from the literature, prior surveys, and publications of statistics.

There are several sources of data that a technical or market researcher can use; some of these are (see also Exhibit 3.1):

- Government sources

  U.S. Census of Manufacturers: useful for statistics.
  U.S. Department of Commerce: useful for statistics.
  U. S. government laboratories: especially useful in areas of expertise.

- Private sources

  Charles H. Kline & Co., Inc.: useful for encyclopedic listings of surveys and services.[159]
  McGraw-Hill Publishing Company: useful for statistics and forecasts.
  Predicasts and similar organizations: useful for statistics and for

surveys of special markets. (Surveys are useful in the ideation phase but are not a substitute for focused market studies.)

Trade and professional associations:[160] useful for statistics and position papers. Source of recommendations on consultants.

Industrial companies: market need, product characteristics, willingness to buy, and price are but a few of the topics on which market researchers can obtain data from knowledgeable people in these firms.

The market researcher might be an individual consultant or a member of a group of market-wise people in a contract research organization. The individual is especially useful for the quick and dirty study done early in the R&D, although in-depth studies can be done by individuals, as well. A market researcher's value depends on personal experience and the extent of associations made in the industry being surveyed. The contract research organization (profit or nonprofit) is most useful for the in-depth study, although one may be contracted for a lesser study, especially if there is follow-up with additional studies.

Many events during R&D can change the validity of a market research effort. Market research may suggest alteration of design involving major changes in the product and hence in the market data. After you have developed a model or prototype, you can test users' reactions to the product and uncover possible technical or marketing problems before major funds are committed to the follow-on phases. As development proceeds, you may find that the product will not sell for the price initially anticipated, and this requires a revision of the original estimate of market potential. You may also discover competition starting to market a better product, possibly using a newer technology, or that new materials or manufacturing processes will change manufacturing costs, thereby permitting or requiring repricing.

A viable product is a marketable product.

It is obvious, therefore, that from the time of the initial market research until the product finally reaches the market, the ball game may change substantially. *Market research must continue along with product development.* Unfortunately, too often management does not recognize that initial market findings are fragile and are time-dependent.

Successful launching of a new product requires not only verifiable market estimates but also the in-house capability to sell the product. In the second stage described above, a question was aimed at determining whether the company has relevant marketing experience, a point often overlooked by the technical person. Consider a company selling widget A into one part of the widget market. Other companies are selling widget B into another part of this market. You, the R&D group, conceive widget C which would replace widget B. You propose it to your company, noting that widget C has several product advantages, not the least of which is that one widget C will replace several widget Bs. The marketing group, however, tells you that they don't have the expertise to sell widget C because the company doesn't sell widget B. Experience with widget A is not enough. To market the product, the company would have to acquire new expertise or make an acquisition that had the necessary expertise, or as an alternative, sell or license the new product to the competition. In the midst of such talk, widget C probably dies.

Another marketing consideration is that a well thought through marketing plan should be derived during the engineering phase; it should show an understanding of the product and the customer and should show, specifically, how the company is going to make money. If marketing cannot derive such a plan, you (the R&D manager) will have wasted your efforts, because the product will never get to the marketplace.

Does marketing believe the product can be sold?

Patents and Licensing

Protect the innovation

Every company does its best to protect proprietary developments and its competitive position through patents and " know how." Managers must therefore appreciate the procedures involved in protecting the basic features of an invention. They must also balance the need to protect proprietary matters with an evaluation of the value of such patents, for patents are not without cost—first and continuing annual costs.

Understand patent features.

Patents have force only in countries where they are granted, so an inventor must file a patent application in every country where such protection is thought necessary. Consideration of foreign filings must be made concomitantly with U.S. filings, and foreign filings must be made within one year of the U.S. filing date in accordance with the International Convention. It is important that you give some early thought to the number of possible patents on a single product and how they interrelate. As you draft a patent application, look at the claims as they define the bounds of the disclosure and study their compatibility with other relevant applications or patents. In one case, a company filed for patents in the United States and in several foreign countries. Then, legal counsel decided to abandon the U.S. application and refile using new findings; however, one of the foreign applications had already issued and now had to be dealt with as prior art. So, this is a tricky arena, and you must act carefully.

If an idea is evaluated as feasible and goes into a full project study, you may find it desirable to file for patent application. You may have already filed an idea record form with your idea review committee and with your company patent department. The next step is to fill out a certificate of invention (or patent disclosure form and a form used for identification of prior art), review these forms with your company's legal counsel, and discuss the merits of filing. Typical forms are shown in Exhibits 8.2 and 8.3.

Carefully consider the merits of filing.

If an idea for a process or product is received from an inventor outside the company, *it is important that you have from the inventor an agreement which certifies that you did not receive the idea in confidence.* It would take an exceptional circumstance to veer from this practice, and then only with legal guidance.

For U.S. utility patents (which cover products and processes) and plant patents, the current effective term is 17 years from the date of patent issuance. For design patents, the term is 14 years. My discussion here is limited to utility patents; please refer to the "References" and "Suggested Readings" sections at the end of this chapter for other discussions.

A patent application is a document prepared according to a certain format[161,162] and filed in the patent office. It is assumed that one normally skilled in the art will understand your patent. The first page of the patent includes information about the inventor(s), the assignee, the date filed, the application number, the field of search, references, and the examiner and attorney, and it presents an abstract and a characteristic drawing. The next pages contain the detailed drawings. The specification, or disclosure, follows and contains the inventor's teachings of his or her invention. More specifically, the disclosure contains a statement giving the field of invention, discussion of prior art, objects of the invention, summary of the invention, elaboration of the invention including a description of the drawings, statements on usefulness, and examples. The application concludes with a list of claims that define the scope of the invention.

The United States Patent Office publishes the *Official Gazette,* a weekly publication that contains a selected drawing and a represen-

PATENT DISCLOSURE

Date:

Please fill in all blanks. If not applicable, say so.

1. Inventor(s) ______________________________
   ____
   ____
2. Title of invention ______________________________
3. Statement of purpose and probable uses:
4. Describe invention:
5. Distinctive features of invention:
6. Information about the invention:
   a. First thought of on or about ______________________________
   b. First explained to ______________________________
   c. First recorded on ______________________________
   d. Constructed on or about ______________________________
   e. Tested on or about ______________________________
   f. Has it been used experimentally? ______________________________
   g. Has it been sold as is or in another product? ______________________________
   h. Have the results been published? ____________
      For sales, please list customers and dates of sale on attached sheet.
      For publications, attach copies.
7. Assignment
   In agreement with the provisions of my (our) employee Invention and Assignment Agreement, I (we) assign the entire right, title, and interest in and to the herein described inventions to: ______________________________
   Company Name

Signature(s) of Inventor(s) ______________________________
____

Witnesses:

This detailed description has been read and understood by me (us) as witnessed by my (our) signature(s).

Witness: ______________________________ ____________
Date

Witness: ______________________________ ____________
Date

Exhibit 8.2 Patent disclosure form.

PRIOR ART

Identify prior art of which you are aware from either internal or external sources including prior patents, publications, sales, offers of sale, uses, etc., and attach copies where appropriate. Doubts should be resolved in favor of identification.

	Information	Relevant Portion
1.		

In preparing the above, I (we) have also sought such information from others likely to be aware of prior art including ( ) supervisor, ( ) colleagues, ( ) others. If checked, identify person(s) contacted.

____________________ ____________________

____________________ ____________________

Exhibit 8.3 Prior art form.

tative claim of each patent granted during the week. Printed copies of patents may be purchased from the patent office or obtained through an attorney or service organization. This is a journal that warrants monitoring by managers of R&D.

While a patent application is being processed, the effective coverage of the claims is negotiated with the patent office. The most common rejection by the patent office consists of prior art that alone or in combination appears to disclose essential features of the applicant's invention as defined by the pending claims. Sometimes the art cited contains such a complete anticipation of the invention that the applicant must abandon it. Usually, however, the applicant can respond by amending his or her application within three months from the mailing of the rejection to receipt of the response by the patent office. Let me warn you that you can expect some creativity on the part of the examiner; he or she may warp the content of another patent or of a combination of other patents to show identity with your invention. Be calm. If warranted, you usually will be able to counter such arguments with discussion of the fundamental science controlling the invention.

Forming part of the amendment is a "Remarks" section in which the applicant attempts to convince the examiner either why he or she was wrong or why, in view of the changes now made to the claims by the applicant, the examiner's previously stated position is no longer valid. In some instances, affidavits or other documentation are used to support arguments. Here is an instance where keeping good records can be most rewarding! If the examiner agrees, the application is allowed and becomes a patent, on payment of the issue fee. If the examiner continues to disagree, the rejection is repeated and made final.

Further amendments are restricted to those that will make the application acceptable without requiring the examiner to make an additional search. A personal interview with the examiner is sometimes helpful in reaching an understanding on what he or she will accept. Failing all else, the final rejection can be appealed to the patent office board of appeals or the application can be abandoned or refiled with new claims. If the application is finally abandoned and not refiled, it will be maintained in confidence in perpetuity by the patent office.

Definitions

Inventions, trademarks, copyrights, and trade secrets are concepts termed *intellectual* property rights. Under the law, all have the characteristics of personal property; that is, they may be bought, sold, licensed, or traded. Trademarks can't be sold except with the business to which they relate.

An *assignment* of a patent and (or) invention is a transfer or sale of the entire rights to the patent and (or) invention to another party.

A *copyright* is an exclusive right to make copies and is granted by the government to authors, composers, or artists or their assigns on works created after January 1, 1978, for their lives plus 50 years. When the work is for hire, the copyright term is for 75 years from the year of first publication or 100 years from the year of its creation, whichever is shorter.

A *disclosure* is a written description of the invention. The disclosure should state how the invention is constructed and operates and should include the signature of a person other than the inventor who, by witnessing the disclosure, shows that he or she has read and understood the invention.

An *infringement* is the unauthorized making, using, or selling of an invention covered by the valid claim of a patent during the life of the patent.

Invention refers to a concept or thing evolved from the mind. An invention may or may not be patentable.

A *license* is a contract between the inventor or the owner of the patent (licensor) and another party (licensee) under which the licensee receives or is granted the right to practice the patent.

An *exclusive license* usually grants all or a portion of the rights under the patent to a licensee without the right of the patent holder to grant the same rights to additional licensees. The licensor or patent owner cannot practice such granted rights unless specifically provided for in the license.

A *nonexclusive license* generally conveys all or a portion of the rights under the patent to another, with the licensor having the right to convey similar or the same rights to other licensees.

A *patent* is a contract between the inventor and the government in which the government grants exclusive right to the inventor to prohibit others from making, using, or selling the invention for a limited period of time. The consideration for this grant is the inventor's disclosure of his or her invention.

A *trademark* is a word, name, symbol, device, or a combination of these adopted and used by a manufacturer or business to identify and distinguish its goods from those manufactured or sold by others.

A *trade secret* is a product or process which is protected against disclosure to the public, thereby giving its owner an advantage over competition.

Licensing or sale

Licensing presents opportunities for commercial gain by obtaining the results of another company's R&D investment. This route to new products reduces the time necessary to get to the marketplace, assuming a marketing capability exists. Licensing of one's own innovations is also a valuable company strategy.

A patent owner may desire to license or sell an invention rather than sell the product covered by the patent, because he or she lacks capital; considers the risks too great, the projected return too low, the invention not in his or her current line of business; or considers that the patent hasn't had the extent of internal use for which it was originally anticipated. Licensing or selling the invention can be considered during or after the ideation phase; however, it is very difficult to consummate a deal until a patent application has been filed and a working prototype has been obtained.

Consider licensing unused patents.

Much of the legwork involved in licensing can be done by the manager in whose area the invention was made, working with the company's designated negotiator. Since this duo represents the licensor, I will refer to them as such for sake of brevity. Before contacting prospective licensees or purchasers of the patent, the licensor determines possible royalty rates and asking and floor prices for the package. Companies that market similar products are identified. Where possible, more than one candidate licensee is approached, the person in charge of the division (or portion of the company that is responsible for the product area of interest) is identified, and then that person is contacted directly. Enough nonconfidential information about the invention is disclosed to create interest. If successful in generating interest, the licensor follows up with a written nonconfidential description of the invention and its benefits. Those who are not serious contenders are weeded out, and at this point, only one candidate is considered, who will be given some time limit for deliberations.

Consider why a possible licensee should need your invention.

The next step is for the licensor to send the candidate company confidential information covering technical and market research results under a confidential relationship agreement, especially where foreign patent applications have yet to be filed. A face-to-face meeting should follow, with possibly a demonstration of the product or process. During this meeting, the first discussions of business arrangements occur, and these must be thought out thoroughly in advance.

The candidate may wish to operate any hardware or pilot plant and to study the opportunity over several months or even a year or two, and thus will take an option to buy or license for a negotiated period with an option to renew. Making money from the option is not the primary factor during its negotiation, but the licensor wants to require enough to give the candidate an incentive to reach an early and favorable decision.

Licensing can be done on an exclusive or nonexclusive basis (defined earlier) and with or without territorial or field-of-use limitations.[163]

Licenses are generally in written form and commonly cover one or more patents with or without "know how."

The licensor wants the maximum royalty obtainable, and the licensee wants the maximum profit (or saving) obtainable under the license. For the licensee, royalty payments represent money that would otherwise be part of profit, so an acceptable royalty depends on the proportion of profit it represents rather than on the proportion of sales. A good agreement, or one which is most likely to be successful, is one in which both parties benefit. A poor agreement is one in which one party's zeal to maximize its own interests is to the detriment of the other. The licensor should consider minimum royalties per year as an assurance of revenue and as a means of shedding unproductive licensees. If the minimums are not met, the licensor can cancel. Faced with minimums, a licensee will usually request that any payments be cumulatively credited against earned royalties and that earned royalties in excess of minimums be cumulatively credited against minimums. See also Udell and Potter.[164]

Many bases exist for royalty determination.

Everyone wins or it's a bad deal.

A licensor may provide incentive to a licensee by having the royalty scale inversely related to sales volume. Licensees may sometimes suggest this as a way of maximizing profit, since the greater the volume of units sold, the more likely the unit cost is to decrease. Also, the licensee often will try to have a cumulatively diminishing royalty, for example, 5 percent for the first 2 million units sold during the term of the license, 4 percent for the next 2 million, 3 percent for the next 2 million, and so on.

From a licensee's viewpoint, it is desirable to limit total royalties to be paid under a license. For example, the licensee can insist on a provision stating that when the amount of royalty paid equals x dollars, the license becomes paid up. The licensee then has an incentive to sell a higher volume more quickly, thus giving the licensor a quicker return and rendering both parties less vulnerable to the inroads of advancing technology.

Disagreement on the unit on which a royalty rate is computed occurs too often. The licensor wants the unit to include as much as possible, and the licensee as little as possible. An example of such a conflict would be a catalyst used in a combustion process. The licensor might want the royalty based on the end-use product, that is, on the combustion system with a catalyst. The prospective licensee would want to pay on the amount of catalyst used. The value of each, and the amount of royalties for each, would be significantly different.

A major problem in some organizations is that licensing or sale of patent rights is left to an ad hoc group or task force. This task force, perhaps only one person, must get on top of technical, market, and legal matters and ferret out potential buyers. When a prospect is found, management wants to get back its investment, everyone gets into the act, logic is lost, recommendations by the task force are ignored, and option and license requirements are designed to which no one will acquiesce. The sale is lost. An important philosophy in licensing and sale is that *everyone must win* for successful consummation and product commercialization.

Licensing should be a formal function.

Summary

- As a technology matures, R&D productivity declines and it becomes harder to make any substantive advance using that particular technology.
- Design screening and evaluation techniques to match the six phases of the innovation process.
- Treat new ideas tenderly; record them, review them, nurture them.
- Establish feasibility through investigation of technical, legal, and market constraints.
- Decision analysis is a most useful technique in the development and evaluation of a project portfolio.
- Market research is an important element of your R&D content from the very beginning of each project.
- Do not accept market research results blindly; always challenge the results and the methods used to obtain them.
- Tailor your product to a selected entry market.
- Protect proprietary developments, but balance this effort with an evaluation of the value of the patents.
- In licensing, everyone must win, or it's a bad deal.

Cases and Incidents

What's the Date of the Invention?

Larry's area of applications development has invented a laser welding process which is being assumed by a division. After some months, the division's technical and legal people descend one day wanting to draft a patent application. The first thing they have to do is draft a new patent disclosure (the first one had not been done adequately). Everyone knows and accepts the inventor, technical and prior art material are easily established, but the date of the invention is in question. A process is different from a product in that it evolves over time, and pinning down an exact date is often difficult.

The lawyers start a review of all pertinent notebooks. Several notebooks by several investigators are reviewed page by page. The lawyers question the investigators as to the exact meaning of their words. Eventually the lawyers and investigators establish the earliest possible date that can be defended by the investigators. Fortunately, the investigators kept good records, that is, they were detailed, legible, and witnessed and dated.

Watts congratulates Shock on his recognition of the importance of well-kept notebooks. Had sloppy notebooks been kept, a date that was defensible could have been much later, giving competition a stronger claim in any subsequent litigation.

When millions of new dollars are at stake, keeping good notebooks becomes a tolerable duty.

Questions

- How do you encourage good record keeping?
- Who follows up on seeing that notebooks are well kept and current?
- Should the keeping of notebooks be included in a person's appraisal?

Will the Real Inventor Please Stand Up?

Using the applications development expertise of his area, Larry starts a cooperative project with a local division making grinding wheels. The new product is to incorporate some of Brown's alloys, a new processing procedure by Larry, and a conceptual design offered by Foster, the division chief engineer. The project goes well. Foster has Oscar, one of his professionals, run some design alternatives through the computer, and Oscar makes some design suggestions on which Foster elaborates. The alloy originally suggested by Brown is changed slightly in composition and content by Roy. Larry's engineer on the task, Tom, eventually works out process refinements to allow the wheel composition to withstand the centrifugal force and give the wear characteristics needed. Harry's electronics crew provides the drive and control system. The components are assembled at the division. Tests are outstanding, and the project is a go.

Enter Andy, the division's patent attorney. He wants to know who is the inventor of this mishmash. He has interviewed a slew of people who've made contributions, and each one claims to be the inventor.

Shock tells Andy he will get back to him. When Andy leaves for another interview, Shock asks Watts what he thinks about the situation. Watts says that Andy must answer his own question! Andy must determine whether someone had the first overall concept (and documented it), and he must determine whose ideas were critical to technical success. Andy must also distinguish new from old art, even though the old art is cleverly applied. He will make his decisions on the basis of notebook reviews and results from his questioning of the investigators. It doesn't matter about the number of investigators on the patent, but they must have made critical contributions. Also, Andy could decide to divide the invention into several inventions made at different times by different inventors.

Shock is happy for Andy to have the responsibility, and very glad the staff's notebooks are in good shape.

Questions

- Who should identify the inventor or inventors?
- Who should explain the selection to the staff?
- How important is documentation to validate the position of each inventor?

Everyone Has to Win

A carryover from early R&D in Click's electronics area is a new electric motor that is touted to be more efficient and less costly than conventional motors of the same size and horsepower range. Shock proposes that the product's patent application and know-how should be sold or licensed. The CEO gives approval. Harry (Click's successor) prepares a prospectus that describes the motor and results of tests in general and nonconfidential terms. Three companies respond, and Shock requests the legal department to prepare a confidential disclosure form.

The legal department refers the matter to the CEO who calls Shock: "Look, Shock, you got permission to determine interest, not to negotiate business deals. I'll take it from here and when I want your help, I'll ask for it. Just remember, you're the technical man; I'm the marketing man."

There is little Shock can say, and he delivers the necessary technical material, a cost breakdown, and a list of benefits.

The CEO, with one of the company lawyers, meets with each of the three firms, and after Harry gives a description of the motor and answers technical questions, the CEO and the lawyer submit their idea of a business deal. The firms say that the CEO's expectations are about 10 times too high, and they each want an option to test the device before any negotiations, providing the CEO's needs can be moderated. They also have some reservations about the technical results. The CEO decides that Shock should go back and do the required experimentation to satisfy their reservations, even though Shock feels these could best be done by the prospective licensee during an option period. The CEO will not consider lowering his demands. Shock tells Watts the story.

Watts is definite: "No one wins unless everybody wins. Your company must compromise, or no deal will be reached, or if one is reached, no product will be produced under your patent. That will mean legal problems."

Harry's crew does some more work, but it is obvious that one of the companies would have done a better and more complete job. The material is transmitted to the CEO, who sends it on to the companies. Two respond as still interested *if* the CEO can negotiate his requirements.

Another meeting is held with the CEO and the companies' lawyers. They finally work out a deal with one company which, although lower, is much higher than the company desired.

The licensee eventually produces a product that it feels is sufficiently different from that licensed as not to warrant a royalty. After much legal conflict, the CEO drops interest.

Watts comments to Shock: "What a mess! You can never expect a royalty to command an excessive portion of the total cost of an item or of the profits that might accrue. As I told you before, you don't have a good deal unless everyone wins. When professionals attempt to reduce a product's cost, that also means a reduction in the royalty."

Shock says, "You know, the boss considers himself a sharp wheeler-dealer but that obviously is not the same as a good negotiator."

Questions

- Should the CEO be directly involved in such negotiations?
- Who should estimate a possible royalty? R&D? Legal? A licensing person? All three?
- Who is best qualified to estimate the reasonableness of the buyer's counter offers? (See the above options.)
- What are the bases for determining a potential royalty?

References

149. R. H. Becker, and L. M. Speltz, "Making More Explicit Forecasts," *Res. Mgmt.*, July–August 1986, pp. 21–23.

150. G. A. Pogany, "Cautions about Using S-Curves," *Res. Mgmt.,* July–August 1986, pp. 24–25.
151. P. A. Roussel, "Technological Maturity Proves a Valid and Important Concept," *Res. Mgmt.,* January–February 1984, pp. 29–34.
152. A. Albala, "Stage Approach for the Evaluation and Selection of R&D Projects," *IEEE Trans. Eng. Mgmt.* **EM-22**(4):153–163 (1975).
153. F. Krawiec, "Evaluating and Selecting Research Projects by Scoring," *Res. Mgmt.,* March–April 1984, pp. 21–25.
154. C. Pacifico and R. Polacek, "New Product Development," Am. Chem. Soc. Audio Course, Washington, D.C., 1981.
155. D. B. Merrifield, *Strategic Analysis, Selection, and Management of R&D Projects,* AMACOM, New York, 1977. Also, "Basic Business Concepts for R&D Management," *Res. Mgmt.,* March 1976, pp. 33–36.
156. L. A. Goldgehn, and R. Lagarce, "Giving a New Dimension to Product Portfolio Analysis," *Mgmt. Rev.,* April 1983, pp. 57–61.
157. C. Freeman et al., "Success and Failure in Industrial Innovation," Centre for the Study of Industrial Innovation, University of Sussex, London, 1972; see also B. C. Twiss, *Managing Technological Innovation,* Longman, London, 1974, p. 17.
158. W. E. Souder, "Promoting an Effective R&D/Marketing Interface," *Res. Mgmt.,* July 1980, pp. 10–15.
159. *Directory of U.S. and Canadian Marketing Surveys and Services,* Charles H. Kline, Fairfield, N.J. (A service updated periodically.)
160. D. S. Akey, *Encyclopedia of Associations,* Gale Research, Detroit, Mich., 1980.
161. J. T. Maynard, *Understanding Chemical Patents,* Am. Chem. Soc., Washington, D.C., 1978.
162. R. W. Mayer (ed.), *Patent Guide for Employees,* Dresser Industries, Dallas, Tex., 1976.
163. D. M. Sell, "Business Aspects of Licensing Patents and Know How," in *Current Developments in Patent Law,* Practicing Law Institute, New York, 1978.
164. G. G. Udell, and T. A. Potter, "Pricing New Technology," *Res. Techn. Mgmt,* July–August 1989, pp. 14–18.

Suggested Readings

Augood, D. R.: "A Review of R&D Evaluation Methods," *IEEE Trans. in Eng. Mgmt.,* **EM-20**(4):114–120 (1973).
Day, G. S.: "Diagnosing the Project Portfolio," *Marketing,* **41**(2):29–38 (1977).
Garcia-Borras, T.: "Research Project Evaluations," *Hydrocarbon Processing,* pt. 1, 1976, pp. 137–146; pt. 2, 1977, pp. 171–182.
McDonald, D. W., and H. S. Leakey: "Licensing Has a Role in Technology Strategic Planning," *Res. Mgmt.,* January–February 1985, pp. 35–40.
Paolini, A., and M. A. Glaser: "Project Selection Methods That Pick Winners," *Res. Mgmt.,* May 1977, pp. 26–29.
Souder, W. E., and T. Mandakovic: "R&D Project Selection Models," *Res. Techn. Mgmt.,* July–August, pp. 36–42.
The Conference Board: "The Marketing Plan," New York, 1981.

Chapter

9

Effective Communication Techniques

Questions answered in Chapter 9

- What are the six means to make communications effective?
- What are the contents of a proposal? of a report?
- What is paramount to assimilation and understanding of a communique?
- What are some rules of telephone usage?
- What is the purpose of any meeting?

You will never be able to carry out a reasonable style of managing without a talent for communications. Nor will you ever fulfill your potential. Margerison[122] says that the effectiveness of a manager depends as much on persuasive communications and leadership as on technical expertise. One of the most important personal attributes you can have is the ability to communicate. You influence people by this ability, you convey progress, you obtain advancement based in great part on this ability, and yet you may pay little attention to improving your communications abilities.

Your success is a function of your communications skills.

Effective communication means that:

- *You are informed:* If you don't have the relevant experience or the necessary facts, whatever you write or say will have little value. Verbosity will not hide a paucity of knowledge.
- *You have planned what you have to say, and you carry out your*

plan: With proper planning, carrying out the plan becomes routine.

- *You have reviewed your presentation or report with colleagues:* Oral presentations practiced before colleagues will help verify completeness and clarity and will build confidence. Practice may not make perfect, but it will help to avoid embarrassments.
- *You have consideration for and a knowledge of your expected audience:* You must recognize the composition of your audience—often a combination of executives, R&D managers, and working scientists and engineers. Prepare your presentation with regard for each; a pure reporting of results is not enough! Each element of your audience must be addressed in terms of area of greatest interest and with regard to ability to understand your message.
- *You recognize the criticality of the first page of a report or proposal or the first few seconds of an oral presentation:* These set the stage for dialogue between your audience and you, and it is here that you may win or lose.
- *You have organized your material so that you can present bite-sized elements:* Simplicity is paramount to assimilation and understanding.

As a manager, you must be expert in communications and train your staff to be as expert as their capabilities allow.

Written Communications

The most formal method of communication is the written word. Recognize the characteristics of your audience, since the several people reading the document will have different corporate responsibilities and different purposes in evaluating its content. Give special attention to the degree of technical sophistication used in each part of the document: direct the summary to management (not highly technical); direct the technical discussion to professionals (highly technical).

The proposal

A proposal conveys the author's need to be funded by his or her audience in order to study a problem or to research an idea. What is to be done must be specifically stated, and the cost justified. Assume in the following that a feasibility review favors proceeding with an R&D project. The proposal should be brief and concise; it has the following content:

- Introduction.

  Introduce and show an understanding of the problem, or introduce and describe in brief your idea and its value.

- Goals.

  State the goals of the immediate study.

- Scope.

  Define the bounds of the effort, what will and will not be done.
  State areas of responsibility.

- Background.

  Provide a clear description of prior studies.
  Compare alternative approaches.
  Show how the proposed approach differs from prior work.
  Include both technical and market information.

- Value of the research.

  State the characteristics of the market that show need for the research.
  State product or process advantages over the competition.
  Indicate what the solution to the problem means to the company in terms of product advantages, profitability, patentability, and so on.

- Method of approach.

  Define the procedures or techniques to be used, and say why they were chosen; relate their use to goals specified earlier.
  Divide the method of approach into phases, giving the milestones to be met; use Gantt charts to illustrate these phases.
  Bring out critical experiments based on identified constraints.

- Qualifications.

  State applicable prior experience of the staff.
  State who will be responsible for specific work.
  Attach one-page bibliographies.
  Describe applicable permanent facilities.

- Reporting.

  State the types of reports that will be submitted, and give the schedule for reporting; mesh this schedule with the milestones listed on the Gantt charts.

- Time and cost.

  Give estimated time and costs for each phase and for the total work. Say that these will not be exceeded without prior approval.

The report

Many professionals view report writing as a useless burden. Report writing can, however, bring order out of chaos, reveal voids in data (by forcing order on a person's thoughts), and instruct on project direction (by forcing one to take time to review what has been and is to be done). As with a proposal, you must consider your audiences and write purposefully to each of them.

Report writing often leads to project clarification.

Good writing begins with organization of your material. Make an outline, and ask if the flow of the content is logical. Only then start writing. The order of your writing is important. First, draft the section (probably an appendix) on methods or techniques used. Then, summarize data into tables and graphs. Draft the experimental results section using the tables and graphs. Discuss the data in a logical, not chronological, sequence. Draft the background section. With these sections written, write the section on discussion of results. Write the introduction, which establishes the intent and goals of the work. The summary and conclusions and the recommendations sections are written last.

In long reports, a highlights section or executive summary at the beginning is mandatory. Place original data in an appendix. Have the report well referenced from the literature and prior art. Some of the report can be written as the work progresses: appendixes, test methods, tables, and figures. This will shorten the time needed for final writing. During the report writing task, other needs of the investigation may emerge.

Be aware that audiences retain what is communicated to a greater extent if more than one of the five senses is affected. The written word is more persuasive if supported by a visual, which also adds another dimension. But remember that visual aids can help or damage depending on how they are prepared[165,166] and how they are used. Add tables, figures, and pictures to your written text.

Technical writing is unique in that paragraphs start with the punch line, not end with it. This way the reader sees your point and then matches what follows to see if it justifies the point made. Simplicity of language and style also promotes clear communication.[167] Keep average sentence length to less than 25 words. Use one of the available computer programs to check spelling and grammar.

Start with your main thought; then explain it.

Remember that any description of poor results should be instructive for future considerations. State negatives in a positive way! For example: "When we combined A with B, we did not get C as expected but, rather, got D. Although disappointed that C was not obtained, we believe that the synthesis offers a way to produce . . ."

Although some may use the personal *I* or *we,* I prefer the impersonal approach in written technical reports and the personal approach in oral presentations. Many prefer the use of the first person in technical writing as being more precise, providing it is done in moderation and with good taste.

Oral Communications

Oral presentations

In a technical report, you may use an impersonal style because the audience is impersonal, that is, not there with you. In an oral presentation, you are on a personal basis and talk to the audience as a friend.

In planning an oral presentation, first develop the theme, narrow down the subject, decide on what you want to accomplish, and so on. Don't have too broad a scope or too many ideas; remember, keep it simple. Next, organize the presentation in logical steps; ask yourself if the audience will follow your organization. Establish your outline. Draft key sentences that you will read, or write out key phrases, but (if possible) do not write your speech. If you must write the speech, veer from it occasionally with some ad lib; look at your audience with only the briefest of references to notes or speech; relax; talk in modulated tones at a moderate speed.

Keep the theme reasonably narrow.

Rehearse the presentation before a mirror, using a tape recorder. A playback may suggest modifications. Timing is essential. Rehearse important speeches in front of colleagues who have different backgrounds, to get their impressions, comments, and suggestions.

During the presentation, use visual aids (tables, figures, pictures), and also pass hardware or chemicals (or simulated chemicals) around or show models. Always coordinate showing slides or passing hardware with the main presentation. Do not let one distract from the other.

Sometimes a slide is self-explanatory, and there is no need to refer to it; at other times you will need to explain it in some detail. In the latter case, you should leave the lectern and point to the slide without blocking any part of it from the audience. This must be done without awkwardness, or else time and the audience interest can be lost.

Look showy with simple visuals, not vice versa.

Overhead projection and slides each have their place. Overhead projection is particularly useful when you wish to draw on the transparency or to use overlays. Slides should not be too busy, and each slide should contain only one idea. Use letters that can be seen from the rear of the room. A rule of thumb is that the distance from the back row of the audience to the screen should never exceed six times the screen width. Machine lettering or machine type produces more legible slides than do typewriters. These same suggestions apply to the overhead projector. Have people in the slides showing equipment. Use colored slides which are self-consistent in lettering and colors. Use graphics rather than complex tables of data.

Remember that the first few seconds or minutes with your boss, a client, or an audience are critical; it is in this short time that you must convince them to be open and responsive to your presentation. Look like you're glad to be there. Have a voice that varies in tone and pitch; a droning voice will lull them to sleep. Be alert and watch the responses of your audience—the smiles, the questioning looks, the frowns (God forbid!). Have a "happy" voice.

Make good during the first few seconds.

Pay attention to your dress. People who look successful and well educated receive preferential treatment in almost all their social or business encounters.[168]

AND MOST IMPORTANT, YOU MUST HAVE SOMETHING OF CONSEQUENCE TO SAY!

I suggest that the following steps are necessary to a good presentation:

- Establish what you wish to accomplish by the presentation: your goal.
- Know your subject.
- Know your audience.

- Consider all possible negative comments and have answers ready.
- Prepare: plan the strategy and tactics of the presentation to attain your goal.
- Communicate to as many senses as possible; the better the information is retained, the better your chances for favorable action.
- Be confident but not arrogant.
- Establish a beachhead: your opening must open their minds to what you say next.
- Tell them quickly your purpose and what you want to say—then say it.

Use the telephone properly

This has been said to be a very trivial subject; however, with either guts or stubbornness, I'm going to persevere and make a few comments. This is the first line of communication and it is often misused by all of us—from the technician to the CEO.

There are two major abuses: The first is not being on the phone when the called party answers. I once called the president of a company but for some reason was not on the line when he answered. He told the secretary abruptly, "If he doesn't think this call important by being on the line when I answer, then I don't feel I need to wait for him," and down went his phone. I still get perturbed when I answer the phone and hear a secretary say, "Mr. E., please hold for Mr. D."

The second abuse is not identifying yourself properly. For example, answering "yeah" rather than "Jones" is plainly rude and does not promote communication—in fact, it is a turnoff.

Rules of answering a telephone include answering the phone promptly and identifying yourself. If you must put the caller on hold, ask permission or give the alternative of your calling back.[169] Speak calmly and get your message across in a concise fashion. Lastly, *please* don't cough, sniffle, wheeze, gurgle, or convey any possibly offensive sound over the telephone.

Meetings

In Chap. 5, I identified the types of meetings that are of value for an R&D group. No other form of communication so easily clarifies a situation, suggests answers to problems, or reveals another's viewpoint. If another's viewpoint is useful, it can immediately be taken advantage of, or if erroneous, it can immediately be corrected in a positive and encouraging manner.

Meetings promote communications, not people.

When you are having a meeting with anyone, especially your boss, plan it in detail. Know what you want to say and what questions you want to ask. Do this quickly and leave. You must be concerned about the other person's available time. However, be certain that you reach an understanding and an agreement, or that the door is left open for further discussions before you leave. If an agreement has special importance, a memorandum confirming it is advisable. This is especially true if several people participate in the meeting. There is a game in which one person whispers a sentence to another, he or she whispers it to another, and so on. The last person discloses the message and it's *never* what the originator said. A confirming memo saves time, tempers, and trouble.

Summary

- Your effectiveness as a manager depends in part on the persuasiveness of your communications.
- By your ability to communicate, you convey progress to colleagues, superiors, and customers; obtain advancement for yourself and your staff; and influence people to collaborate and to be productive.
- In any type of presentation, know your audience, plan, have a narrow subject, keep it simple, use visual aids, and influence your audience favorably with your opening statements.
- Use the telephone properly and influence people appropriately.
- Meetings are to promote communications, not people.

Cases and Incidents

Don't Tell Them, Communicate with Them

June is a chemist in Brown's fundamental alloy investigation area and has an interest in technical forecasting. This leads her to take a course on forecasting and to do some analyses on certain of the company's products. One chemical is produced using process A by the company and using process B by many of the competitors. There are some differences in the quality of the product produced by each process, and many companies have been switching over to process B during the last decade.

June does a Fisher-Pry analysis and finds that the chemical produced by process A is losing its market share to that produced by process B and will be out of use in the next decade. This information is passed to Brown, to Shock, to the CEO, to the president of the division, and to the division planner. The answer goes the reverse route and essentially says that the Fisher-Pry method is not applicable and that company forecasts are backed up by trade association forecasts showing continued sales by both processes into the foreseeable future.

Shock, talking to Watts, says that if the forecast by June has any validity, it should have forced some introspection. What went wrong?

Watts suggests that two errors were made. First, no single forecasting method has real value until it is backed up by a second method. In this instance, expert opinion would have helped. Second, the manner of getting the result to the division piqued the planner, and he reacted defensively. It is a human reaction. It would have been best to communicate informally with the division and then to send them the results directly. You don't want to appear to be telling them what their business is.

Questions

- What other method could have been used to confirm or deny the Fisher-Pry result?
- Should someone from the R&D group have called the division and discussed the forecast informally?
- Should the memo have gone upstairs?
- Were you the planner, what would you have done?

Take Six Months—If You Have To!

New corporate management brings a change in management of the R&D group. Further, management embarks on an across-the-board belt-tightening and reorganization. Some businesses go on the block, others are acquired, personnel are culled, and many functions are limited or eliminated. The push is on to reduce costs and improve the bottom line, much of which was long overdue. However, there is the fear that a general climate is beginning to form of short-term efforts yielding immediate profits prevailing over long-term efforts to enhance profitability through new products.

Larry gets the idea for a new design of an existing product based on use of Brown's alloys. If the idea works, it will revolutionize the business. The general concept is presented at a meeting and the management jumps on it, because they also can see the potential. Larry is told, "Take six months; we want that product; it will help us to be leaders in the field."

Shock is taken aback by the statement and mutters something to the effect that it could take a bit longer; however, no real dialogue ensues.

Longer, indeed! It takes six months just to design the product.

Now each component has to be made. Processing and fabrication procedures have to be worked out. Assembly procedures. Quality control. A year into these, and other problems of applying the coating and maintaining tolerances seem to be stumbling blocks. There is a period of trial and error in which no significant progress can be reported. Then progress is finally made, and it seems that all is a go, when an order comes down to cancel the effort.

On checking back with management, Shock finds that word had been obtained from the division that would produce the new product that it could not be produced at a competitive price. They were also skeptical that any product would be practical, although the technical work so far had been exemplary. The economic analysis was countered by one from the R&D group, but who was management to believe? The project was terminated.

Shock tells Watts that he is very disappointed, that this was a good project, and he firmly believes that a competitive product would have evolved.

Watts is sympathetic but says that the case is rather more complex than might be apparent. First, management did not recognize the amount of R&D needed; second, a long-term project with its attendant initial low probability of success is not easily fundable under present circumstances; third, the division appears intent on a higher margin of profit from a new product than might be customary. The combination is almost unbeatable. Perhaps Shock lost this one at the starting gate. Before starting the effort, the initial project planning should have been documented and presented to corporate and operations to obtain agreement. The constraints of management and the division could have been reconciled or the concept terminated without the subsequent expenditure of resources.

Shock is glum. When is he going to learn everything he needs to know for every possible situation?

Questions

- Should Shock have immediately presented management with a GANTT chart of tasks and times and discussed the project before initiation?
- Should R&D and the division have gotten together at the start and discussed constraints and how the division evaluates results?
- Will this action have a long-term effect on the R&D group?

References

165. "Overhead Projection Tips," 3M Company, Minneapolis, Minn., 1960.
166. "More Effective Meetings," 3M Company, Minneapolis, Minn. (no date).
167. R. Flesch, *The Art of Plain Talk,* Harper, New York, 1946.
168. J. T. Molloy, *Dress for Success,* Wyden, New York, 1975.
169. Pacific Telephone: "You and Your Telephone" (no date).

Suggested Readings

Wismer, R. D.: "Making Effective Technical Presentations," *Res. Techn. Mgmt.,* July–August, 1989, p. 9.

Chapter

10

Measuring Profitability as One Control Activity

Questions answered in Chapter 10

- What are the various ways of measuring profitability?
- What is return on investment?
- What is payback period?
- What is net present value?
- What is internal rate of return?
- What are some uses of profitability measures?
- Why is market share important, and how can it be estimated?

In prior sections, I discussed the need for managers to review projects, plans, and the performance of staff as part of their control activities. One of the most important of these control activities is monitoring profitability. Various means to measure profitability are presented here.

Enhancing profitability is what it's all about! Maximizing over time a company's profit over and above the capital invested! Managers and senior professionals must be capable of calculating *return on investment* (ROI), *payback period, internal rate of return* (IRR), *net present value* (NPV), or *venture worth,* and *market share.*

Profitability can be expressed in terms of several measures.

Many professionals balk at the mere thought of controls; even some managers are hesitant to enforce controls. Yet, control of any activity is necessary if it is to be successful. As a manager, you control your area or group by monitoring and reviewing its output. Your managers or senior professionals must calculate profitability measures, and it is your prime responsibility as manager to see that these are done routinely and submitted to you for review and approval. Knowledge is control; control enhances success.

A successful activity is a controlled activity.

Measures of Profitability

ROI, an important measure of profitability,[170] relates investment and performance quantitatively. Whether determining if a new product may be profitable or requisitioning a new piece of equipment, you will be calculating ROI. Return is income earned on a given investment. Income denotes profit, *earnings before taxes* (EBT), *earnings after taxes* (EAT), or net earnings. Investment is the resource dedicated to a project or the purchase of security with the expectation of a gain. ROI is, therefore, the ratio of two values. In terms of your personal finances, if you place $1000 in a savings account and realize $100 interest in one year, your ROI is 10 percent before taxes and $100 is your EBT.

Be certain you know what you mean when you use these terms. For example, ROI can be based on initial investment or average investment, on EBT or EAT, or on a discounted cash flow before or after taxes.

If the *present value* (PV) of all cash outflows is equated to the PV of all cash inflows by a discount rate, that rate of return has been variously called the *internal rate of return* (IRR), *interest rate of return, profitability index,* or the *investor's method.*[171] Rather than stipulate a rate of return that you must realize, solve for the discount rate and compare it with a risk-adjusted return.[172] This is an important measure when relating how competitive a new product may be with that of a competitor.

Another measure of profitability is the payback period. Calculate it from the present value of the cash flow and the original investment with variations such as whether the cash flow is measured before or after taxes. Your company may have a rigid payback period which could affect the type of R&D that you can entertain.

In the venture worth method, you discount cash inflows and cash outflows at some minimum acceptable return on capital. The NPV is then the PV of the inflow less the PV of the outflow. Refer to portfolio analysis and Table 7.5 in Chap. 7.

Other Inputs to the Calculations

Working capital corresponds to the funds needed to get a project going and to meet such obligations as inventories, accounts receivable, cash for materials, and so on. *Accounts receivable* are obligations owed to a company by customers for goods or services received. Accounts receivable, like inventories, should be kept to a minimum.

Depreciation is an allowable expense; except for land, all capital assets have a finite life. If you depreciate something over time, you can obtain your depreciation expense by the straight-line method:

$$\frac{\text{Value of item} - \text{salvage value}}{\text{Years}} = \text{depreciation}$$

Depreciation expense reduces reported profit but does not entail a cash outflow. It is a real charge and can become a significant part of your budget. Thus you must be assured that new capital assets are needed before purchase and that they will receive full use after purchase. As you will see later, although depreciation reduces profit, it increases cash flow, so cash flow and profit can be significantly different.

Besides straight-line depreciation, you can recognize that reduction in market value is greatest in early years and use sum-of-the-years digits and double-declining-balance methods. Assume an asset is depreciated over five years. You multiply the book value by a fraction that declines each year. To get the fraction's denominator, add 5 + 4 + 3 + 2 + 1 = 15; the fraction by years is then 5/15, 4/15, 3/15, 2/15, and 1/15. Compute double-declining-balance depreciation by doubling the straight-line depreciation charges computed for the remaining book value of the asset.

Managers and senior professionals must also calculate the time value of money.[172] It becomes necessary to make these calculations in instances such as portfolio analysis discussed in Chap. 8.

$$\text{FV} = \text{PV} + \text{PV} \times i = \text{PV}(1 + i)$$

where FV = future value of money
PV = present value of money
i = interest rate as a fraction

$$\text{FV}_1 = \text{PV}(1 + i) \qquad \text{end of first year}$$

$$\text{FV}_2 = \text{FV}_1 + \text{FV}_1 \times i \qquad \text{end of second year}$$

$$= \text{FV}_1 (1 + i)$$

$$= \text{PV}(1 + i)(1 + i)$$

$$\text{FV}_2 = \text{PV}(1 + i)^2$$

Generalizing

$$FV_n = PV(1 + i)^n$$

This exercise gives you the sum you might be willing to accept now in lieu of a promise to receive a larger sum in the future. You would accept this smaller sum if you could get at least the interest rate on the PV that would mathematically yield the future value. Thus, you would solve for PV in the preceding equation:

$$PV = FV_n \frac{1}{(1 + i)^n}$$

PV, then, is the future value FV_n times a fraction. These fractions are shown in Table 10.1 for use in PV calculations.

Profitability Criteria

Every company has its own definition of minimum acceptable values of profitability For example, ROI on an EBT basis could be stipulated to be 35 percent or higher. For project analysis, ROI on an EBT and a discounted cash flow (DCF) basis is useful. There are special occasions when a lower or even negative ROI may be acceptable, but everyone must know it is negative and what its eventual disposition will be.

IRR on an after-tax basis should be greater than the current cost of money. Ziemer[173] suggests that the value of IRR that should be sought can be estimated by the sum of long-term interest rates and the business risk; for example, IRR = interest + $(1 - P)$ where $(1 - P)$ is risk and where P is the probability of technical success. So, if the interest is 15 percent and the investment has a 90 percent probability of success, the IRR target is $0.15 + (1 - 0.9) = 0.25$, or 25 percent.

One way to assess the acceptability of a new product to a prospective customer and hence the product's marketability is to relate the internal rate of return for the new product to that for the product the customer currently uses. A differential rate of return may also be useful. For example, if the customer is buying product X and you wish to sell product Y, subtract the X investment and cash flows from the corresponding Y values and recalculate an IRR on the incremental values. The resulting IRR should be greater than 20 percent to be interesting.

Payback periods of less than three years are good, and periods greater than five years are poor.

The NPV should be positive to make a project acceptable.

TABLE 10.1 Table of Present Values

	Discount rate																	
Year	5	10	15	20	25	30	35	40	45	50	55	60	65	70	75	80	85	90	Year
1	.9524	.9001	.8696	.8333	.8000	.7692	.7407	.7143	.6900	.6667	.6452	.6250	.6061	.5882	.5714	.5556	.5405	.5263	1
2	9070	.8264	.7561	.6944	.6400	.5917	.5487	.5102	.4756	.4444	.4162	.3906	.3673	.3460	.3265	.3086	.2922	.2770	2
3	.8638	.7513	.6575	.5787	.5120	.4552	.4064	.3644	.3280	.2963	.2685	.2441	.2226	.2035	.1866	.1715	.1573	.1458	3
4	.8227	.6830	.5718	.4823	.4096	.3501	.3011	.2603	.2262	.1975	.1732	.1526	.1349	.1197	.1066	.0953	.0854	.0767	4
5	.7835	.6209	.4972	.4019	.3277	.2693	.2230	.1859	.1560	.1317	.1118	.0954	.0818	.0704	.0609	.0529	.0461	.0404	5
6	.7462	.5045	.4323	.3349	.2621	.2072	.1352	.1328	.1076	.0878	.0721	.0596	.0496	.0414	.0348	.0234	.0249	.0213	6
7	.7107	.5132	.3759	.2791	.2097	.1594	.1224	.0949	.0742	.0585	.0465	.0373	.0300	.0244	.0199	.0163	.0135	.0112	7
8	.6768	.4665	.3269	.2326	.1678	.1226	.0906	.0678	.0512	.0390	.0300	.0233	.0182	.0143	.0114	.0091	.0073	.0059	8
9	.6446	.4241	.2843	.1938	.1342	.0943	.0671	.0484	.0353	.0260	.0194	.0146	.0110	.0084	.0065	.0050	.0039	.0031	9
10	.6139	.3855	.2472	.1615	.1074	.0725	.0437	.0345	.0243	.0173	.0125	.0031	.0067	.0050	.0037	.0028	.0021	.0016	10
11	.5847	.3505	.2149	.1346	.0859	.0558	.0368	.0247	.0168	.0116	.0081	.0057	.0041	.0029	.0021	.0016	.0012	.0009	11
12	.5568	.3186	.1869	.1122	.0687	.0429	.0273	.0176	.0116	.0077	.0052	.0036	.0025	.0017	.0012	.0009	.0006	.0005	12
13	.5303	.2897	.1625	.0935	.0550	.0330	.0202	.0126	.0080	.0051	.0034	.0022	.0015	.0010	.0007	—	—	—	13
14	.5051	.2633	.1413	.0779	.0440	.0254	.0150	.0090	.0055	.0034	.0022	.0014	.0003	.0006	—	—	—	—	14
15	.4810	.2394	.1229	.0649	.0352	.0195	.0111	.0064	.0038	.0023	.0014	.0009	.0005	—	—	—	—	—	15

Use of Profitability Measures

I have set up a hypothetical situation in Table 10.2 so that you can see what the various measures are. Subtract from sales the variable costs. These are costs which vary in direct proportion to the quantity of products produced, such as materials, labor, and variable overhead. The result is variable margin. Subtract fixed costs. These are costs that are not affected by product quantity, such as indirect plant cost, management expense, and sales expenses. The result is EBT. Now you come to a decision point: should you calculate on EBT or EAT? On investments at the corporate level or within a small company, EAT is the more realistic measure. On projects of subdivisions within a large corporation, calculations using EBT become more practical and this is probably what you should use. Which method or combination of methods you use is chosen by you or, more likely, is dictated by company policy. Let's try them all.

Some of the calculations shown in Table 10.2 may be useful in project evaluations before technology transfers. In these calculations I assumed a 50 percent tax bracket in calculating an income tax (obviously a high number, but easy to use); subtraction gives EAT. After determining EAT, I computed investment by year. The initial investment was noted and additional investments were entered for later years. Working capital was entered and, in this instance, I arbitrarily used 30 percent of each annual incremental increase in sales. Enter depreciation as a negative value, using your method or your company's method; I used the straight-line method here. Return of capital is conventional in the last year of the project. The investment at the end of the period was entered and became the investment at the start of the next period. The average investment is the numerical average. This method was continued to the end of the project, in this instance, the tenth year.

I then calculated cash flow by year. EBT, EAT, and depreciation were reentered for convenience. Other cash items were entered, in this instance, the working capital as negative values. Residual value is the net assets in the last year. Net cash flow before and after taxes was then calculated.

To obtain discounted cash flow, I used the values in Table 10.1 and interpolated between the values to obtain ROIs on original investment.

For initial and additional investments, the internal rates of return were calculated using the relationships noted above in a computer program.

I obtained cumulative cash flows, and from these estimated the payback period. Various authors differ in their methods for calculat-

TABLE 10.2 **Project Analysis ($000)**

Item	Year 1	2	3	4	5	6	7	8	9	10	Total	Comments
1. Sales	3000	7000	15000	30000	60000	100000	100000	100000	100000	100000	615000	
2. Variable costs (materials, labor, var. O/H)	2000	4000	7000	16000	28000	45000	45000	45000	45000	45000	282000	
3. Variable margin (1–2)	1000	3000	8000	14000	32000	55000	55000	55000	55000	55000	333000	
4. Fixed costs (indirect plant costs, management, and selling expenses)	1200	2000	4200	8200	15000	26000	26000	26000	26000	26000	160600	
5. Earnings before taxes (3–4)	-200	1000	3800	5800	17000	29000	29000	29000	29000	29000	172400	
6. Income tax	-100	500	1900	2900	8500	14500	14500	14500	14500	14500		Used 0.5 factor
7. Earnings after taxes (5–6)	-100	500	1900	2900	8500	14500	14500	14500	14500	14500		
8. Investment by year												
(*a*) Investment, beginning of period	12300	12785	13655	15695	19805	28115	39425	39735	38045	37355	255915	
(*b*) Additional investments	—	300	300	300	—	—	—	—	—	—	900	
(*c*) Working capital	1100	1200	2400	4500	9000	12000	—	—	—	—	30200	
(*d*) Depreciation	-615	-630	-660	-690	-690	-690	-690	-690	-690	-690	-6735	
(*e*) Return of capital	—	—	—	—	—	—	—	—	—	-36665	-36665	
(*f*) Investment, end of period	12785	13655	15695	19805	28115	39425	38735	38045	37355	—	-243975	
9. Cash flow by year												
(*a*) Earnings before taxes (5)	-200	1000	3800	5800	17000	29000	29000	29000	29000	29000	172400	
(*b*) Earnings after taxes (7)	-100	500	1900	2900	8500	14500	14500	14500	14500	14500		
(*c*) Depreciation	615	630	660	690	690	690	690	690	690	690	6735	
(*d*) Other cash items	-1100	-1200	-2400	-4500	-9000	-1200	—	—	—	—	-30200	Working capital
(*e*) Residual value (8*e*)	—	—	—	—	—	—	—	—	—	36665		
(*f*) Net cash flow before taxes (9*a*+9*c* + 9*d* + 9*e*)	-685	430	2060	1990	8690	17690	29690	29690	29690	66355	195600	

TABLE 10.2 Project Analysis ($000) (*Continued*)

Item	Year 1	2	3	4	5	6	7	8	9	10	Total	Comments
9. Cash flow by year (cont'd)												
(*g*) Net cash flow after taxes (9*b*+9*c* + 9*d* + 9*e*)	-585	-70	160	-910	190	3190	15190	15190	15190	51855		
10. Discounted cash flow—BT (DF × 9*f*)												DF = discount factors
Trial	-507	236	837	599	1938	2922	3634	2690	1992	3298	17639	35%
Trial	-489	219	751	518	1615	2349	2818	2013	1437	1789	13020	40%
Trial	-473	205	676	450	1356	1903	2203	1560	1048	1312	10500	45%
11. Discounted cash flow —AT (DF × 9*g*)												
Trial	-468	-45	82	-373	62	836	3185	2549	2038	5569	13435	25%
Trial	-450	-41	73	-319	51	661	2421	1862	1432	3759	5449	30%
12. ROI (DCF & BT)	[(13020 − 12300)/(13020 − 10500)]5 + 40 = 41.4%											Initial investment
13. ROI(DCF & AT)	[(13435 − 12300)/(13435 − 9449)]5 +2 5 = 26.4%											Initial investment
14. Internal rate of return —BT; AT	42% 26%											
15. Cumulative cash flow—BT (Σ*f*)	-685	-255	1805	3795	12485	30175	59865	89555	119245	185600		
16. Payback period—BT	(12300 − 3795)/(12485 − 3795) = 0.98, so ~ 5 y											
17. Cumulative cash flow —AT (Σ9*g*)	-585	-655	-495	-1405	-1215	1975	17615	32355	47545	99400		
18. Payback period—AT	(12300 − 1975)/(17165 − 1975) = 0.68, or ~ 6.7 y											
19. Venture worth—after tax at 15%												
20. Payback period—using line 5	1.4 y											

ing payback. Rosenau[174] calculates it as illustrated in step 18 of the table. Jelen[171] and Pacifico and Polacek[154] calculate this measure before adding depreciation, and Polacek suggests before taxes and depreciation. Their approach would use EBT (line 5 of the table), which gives a value of 1.4 years, significantly different from the other values in the table. You must, of course, adjust reference values to what your company is using. I suggest starting with the method used in line 16 of the table, unless your company stipulates otherwise.

The last task is to calculate the NPV, or venture worth, using 15 percent as the cost of money.

Now, let's recapitulate: How do our results compare with our ballpark minimums? ROI: EBT is 41 percent versus the 35 percent minimum. IRR: AT is 26 percent versus the 25 percent for the hypothetical interest rate plus risk factor. The payback period is 5 to 7 years, which is poor, and if the project proceeds, something needs to be done to enhance cash flows in the early years. Venture worth, or NPV, is positive. So, all measures look good except for the payback period.

Know when you have competitive strength.

As the R&D manager, you need at some point to estimate the sales price of a competitive product. A quick estimate can be obtained using a value of variable costs provided by the manufacturing expert:

$$\text{Price} = \frac{\text{variable costs}}{1 - \text{variable margin}}$$

Variable margin (VM) should be at least 0.5, or 50 percent, for a new product, and if you can compete at 70 percent VM, you've got economic strength.

An elegant way of calculating sales price has been shown by Herbert and Bisio[175] using ROI and DCF calculations. Rondeau[176] gets at ballpark estimates from materials, costs, weight, and wastage and tooling estimates. He multiplies these by a number that reflects whether the product is made in house or outside and, finally, by an experience factor which is a percent of sales.

Requests for appropriations for capital investments (other than legally required for such items as emission controls) from your R&D area or group may use all or one of these measures. ROI can also be used in estimating the worth of acquisitions.[170] The net cash flow over a 10-year period is calculated from the sum of EAT and depreciation less working capital. These annual values are then discounted at some acceptable percentage. The sum of these discounted cash flows plus return of the working capital approximates the value of the acquisition.

Since ROI is a function of market share,[16] you can use market share as a profitability measure. Estimating the market share that a new product might capture can also be used in project selection and in market research studies. In the section on "Substitution Technology" in Chap. 7, I discussed a method for estimating market share and the mathematical procedures to be used as adjuncts (Fisher-Pry and Blackman's methods).

Pacifico and Polacek[154] have suggested another means of estimating market share for new products from the approximate equation:

$$\text{Market share (\$)} = \text{sales}\,\frac{\text{DF}}{2(n + 1)F}$$

where F = market availability factor (fraction of market available)
n = number of suppliers
DF = $(A \times B)/C$
A = net advantages
B = probability of achievment
C = customer investment required

Summary

- Enhancing profitability is what it's all about. To enhance it, you must measure it. You measure profitability by calculating return on investment, payback period, net present value, or internal rate of return.
- *Return on investment* (ROI) is the income earned on a given investment in a given year.
- *Payback period* is the elapsed time to recover the initial investment.
- *Net present value* (NPV) is the sum of discounted cash flows.
- *Internal rate of return* (IRR) is the discount rate that equates the present values of expected cash outflows and inflows.
- Since ROI is a function of market share, market share can be used as an indicator of profitability.

Cases and Incidents

If You Don't Buy a Ticket

Harry, manager of the electronics area, is not getting new projects started as readily as Shock thinks he should. Shock finds that there are always plenty of ideas in his area but Harry always has some reservations about technical success.

Shock says to Harry, "You must decide on the technical virtues of the ideas, and you shouldn't waste money on obviously fruitless ventures, but you must also consider everyone's opinion and give others in your area a chance to work on their ideas. Remember, if you don't buy a ticket, you don't win at the races. And by giving others a chance, you will encourage creativity."

Harry says, "Yeah, but what if you bet on the wrong horse?" Seeing Shock's expression, he quickly adds, "OK, we may waste some time and money, but I'll give it a whirl."

How Are They Going to Know Me from a Plumber?

Elsie is an exceptionally good chemist and wants new business cards printed as Elsie M. Smith, Ph.D. Shock authorizes the cards but without the Ph.D.

Elsie asks, "How are they going to know me from a plumber?"

Shock replies, "By your wisdom and excellence and the big words you use."

The Egoist

Elsie rehearses her annual presentation to corporate management in front of the area managers. It is an outgoing, enthusiastic presentation consuming about twice the allotted time. The general consensus is that she has done a better job selling herself than in communicating her subject.

Watts comments, "Remember the old acronym, *KISS* (keep it simple, stupid)."

Smiley

You're probably too young to remember Smiley, but he was a doom and gloom guy that was always giving some grim announcement or prediction. Well, Larry had a professional who spent every free moment criticizing management, local and corporate; the idiots never did anything right. However, he was a good investigator when you finally got him to work on a project. The problem was to convince him, first, that the project had value and, second, that he was the best person for the job. Once into the project, he handled it professionally.

This employee's consistently negative approach to any conversation and his adverseness to management soon got him his label of "Smiley." It also got him a transfer.

The Master

Jack is in charge of general maintenance in Shock's group; he is generally recognized for his knowledge, enthusiasm, loyalty, and initiative (don't tell him to do something and later change your mind, because you'll be too late; it will already have been done). However, Jack has his own priorities, and friendly people get service first.

Gerry has to wait an unacceptable length of time to get some work done and calls Jack in to explain to him the facts of life. After 2 hours of patiently listening to the "real" facts (such as Gerry's work is important, probably the most important around here; Gerry is a Ph.D. and expects respect; when he says "jump," jump; Jack has an attitude problem; and so on and so on), Jack calmly gets up, looks Gerry in the eye, and tells him he's a silly ass and he can take his mumbo jumbo and stuff it. With that, he walks out on a stunned Gerry.

References

170. S. R. Goodman, "Expanded Uses of the ROI Concept," *Financial Executive,* March 1968, pp. 28–34.
171. F. C. Jelen, *Cost and Optimization Engineering,* McGraw-Hill, New York, 1970.
172. H. T. Spiro, *Finance for the Nonfinancial Manager,* Wiley, New York, 1977.
173. D. R. Ziemer, "Qualifying Level of Investment in Growth and Renewal," *Long-Range Planning,* August 1976, pp. 2–17.
174. M. D. Rosenau, Jr., *Innovation,* Lifetime Learning Publications, Belmont, Calif., 1982.
175. V. D. Herbert, and A. Bisio, "Venture Analysis, The Risk and the Benefit, part 2," *Chemtech,* July 1976, pp. 423–429.
176. H. F. Rondeau, "The 1-3-9 Rule for Product Cost Estimation," *Mach. Design,* August 21, 1975, pp. 50–53.

Suggested Readings

Stewart, R. D.: *Cost Estimating,* Wiley, New York, 1982.

Postscript

If you have the technical skills of a good manager and are versed in planning, implementing, leading, and controlling, as discussed in the text, can you be an effective and efficient manager? The answer must be *maybe,* because of other circumstances. If you read about specific topics such as entrepreneuring (and I hope you do), will that ensure success? That will be helpful, but the answer is still *maybe.* The hard truth of innovation, as continually brought out in the text, is that it depends on many people carrying out many different activities. Your answer lies, in part, in whether your corporation recognizes the required holistic approach toward managing innovation, which is that all managers in the innovation process must have favorable characteristics and skills, perform their management activities to the best of their potential, and foster collaboration through their management style. The corporation's position determines the freedom or opportunity that the R&D manager has to be effective and efficient. The corporate position also promotes that climate and culture that attracts people and encourages them to be an active part of the innovation scene.

Think both of classical management skills and the network of interpersonal relationships.

Examine your management network, whether manager of an area or of an R&D group. Do you have an innovation management circle? Do all the managers cooperate to make it work? Have you hired a superior staff? Is your R&D portfolio the result of good planning? Are projects being carried out efficiently and effectively? Do you believe that your management style is promoting achievement in your staff? Are you in sync with your boss? Do you have reasonable rapport with support and operating groups? If the answers are "yes," you are probably operating with a reasonable holistic approach and have a good chance of being successful with innovations.

The R&D manager must get everyone rowing in the same direction.

Only a *good chance?* Your chance depends not only on *your* actions but also on the actions of others within the management circle. Refer back to Fig. 2.1 in Chap. 2. If all these role players have an analogous holistic approach, you should be in a state of euphoria, because the only roadblocks will be the intrinsic technical aspects of R&D or problems brought about by you or your staff in handling people. You've got it made!

If, on the other hand, you are having problems obtaining the participation of some people, examine your management style first to see if you are at fault. If you feel that your approaches to others have been appropriate, then you must talk with them, emphasizing the value of their participation to the success of the innovation. Determine if the managers above you in the management hierarchy understand and promote the holistic approach. If not, you could be in trouble. Your productivity may be zilch even with an excellent creative staff, because the needed environment is not promoted from the top down through all levels, especially in the downstream groups responsible for final development and commercialization of new products.

If you are the CEO and you are not getting what you believe the company should from R&D, examine the path from you to the R&D group and from them to the groups responsible for developing the final products. Do you see all the elements of the holistic approach being practiced? You should be able to spot the weak link in the process and make appropriate changes in either conduct or staff. Of course, the weak links can be several; if so, perhaps you need to reflect on your own skills and management style and on those of the people in your staff.

To win with technical innovation, every manager must have good planning techniques, have a competent staff with good consultants to carry out the plans properly, have a management style that influences others to achieve, have the cooperation of all role players in innovation's management circle, and work toward an environment in which everyone recognizes the need for this total approach. The various elements of the holistic approach are set forth below in the form of questions. Answer them with a "yes," and you should have no concern. Any qualification to an answer will be a basis for determination of how you can make improvement or whether improvement is beyond your purview.

1. Do you have a corporate management team that is aware of the many factors influencing success in the innovation process and aware of the team's participatory role?

   Every facet of management affects the degree of innovation existing in a company. The process of innovation is the life process of a company, and there is much literature to support the conclusion that innovation is critical to industrial and national productivity. Successful implementation of R&D requires an involved management.

2. Do you have a corporatewide R&D plan that is integrated with the corporate strategic plan?

   It is a top management duty to see that the corporate environment fosters innovation and that the corporatewide R&D plan is integrated with the corporate strategic plan. R&D will, therefore, create in directions that support the corporate strategic plan. Managers should get a fix on corporate needs, on the company's risk attitude, and on goals desired by corporate. Then the six-stage procedure given in Chap. 3 can be used to derive a strategic technical plan.

3. If you are the head of R&D, do your managers have knowledge of the several activities that lead to successful implementation of the R&D plan?

   To carry out the plan properly, you must give attention to creativity, idea screening, deployment of personnel, and use of consultants. You must be aware of the benefits gained from the use of facilitators. Your staffs must recognize constraints early and have empathy for the needs of downstream groups. Market research must be continually upgraded, and you must always strive to improve communications.

4. Is there a corporatewide environment that encourages needed participation and promotes technology transfers?

   Technology transfer is a much overlooked facet of the innovation process. You can plan all you want, do portfolio analyses, and have sophisticated project selection and evaluation methods, but you will *fail* unless the company stresses promoting *all* phases of innovation. So, there is much beyond the activities of the R&D group in this total innovation process. R&D must obtain early participation of engineering, manufacturing, and marketing. As a project matures and proceeds down the innovation pipeline, management must assure facile technology transfers and coupling of all functions. These might occur naturally and, hence, atypically; a few directives from top management help immeasurably.

5. Does management hire managers who hire other managers who foster the total innovation process? Do these managers have the characteristics, skills, and management approaches that are the basis for a successful management style?

   Top managers must be recruited who will recruit top technical staff. Attitudes of people can be positive, complementing innovation efforts, or negative, working contrary to the company's welfare. Such attitudes are influenced by the management system and a manager's style. Remember that it is how others perceive your characteristics, skills, and management approach that leads to your influence and effectiveness with them.

6. Do managers set high standards with the expectations that they will be met?

   To achieve, a staff must not only have excellent abilities, but also perform up to their potentials in fulfilling goals set for them and the group.

7. Within the corporate policies and procedures, is it possible for managers to reward performance reasonably?

   People vary in their needs not only among themselves but also over time and under every set of circumstances. Managers must recognize this variability in order to provide appropriate rewards on a performance basis.

8. Is it recognized on a corporatewide basis that all the above must be in play? That a holistic approach is the means to optimize success?

   A corporatewide holistic approach means that R&D does not operate in an ivory tower insulated from others in the corporation whose cooperation, or lack of cooperation, may mean success or failure to a new product. It means, therefore, that managers must be aware of the management circle and must communicate with and support the needs of others in it. R&D managers must take the initiative to obtain the participation of others in this circle and, particularly, the enthusiastic follow-through of downstream groups.

If the company approves a reasonable proportion of long-range R&D and generally has the above elements in play, winning depends on you. Also, nothing is perfect, and one shouldn't talk gloom and doom if all the above is not in play. However, if you are in a company where the emphasis is on the quarter ahead or even the year ahead, you must begin to ask whether this is a short-term situation requiring belt-tightening or whether it reflects a real and continuing short-term attitude emanating from the top down. If the latter, I think we have established in the text that this company will never be a leader and must decline. Bite the bullet and find a better company. If, on the other hand, you can work with others in the company toward getting all elements into play, then go to it!

Nothing is perfect, so make allowances; if things look grim, reexamine your goals and opportunities.

Index

Note: Page numbers followed by *n* indicate material in footnotes.

About the Author

Robert D. Englert is a consultant specializing in the management of research and development. Dr. Englert has more than 30 years' experience in research management, including positions as Vice President and General Manager of the Advanced Technology Center at Dresser Industries and Executive Director of the Southern California Laboratories of the Stanford Research Institute (SRI). While at SRI, Dr. Englert directed the expansion of the Institute's science division from a narrow focus on air pollution to wide-ranging research achievements in physical chemistry, organic chemistry, chemical and mechanical engineering, and plant biology. Dr. Englert has written 13 papers and holds 13 patents. He earned his Ph.D. in chemistry at the University of Colorado.